AN ELEMENTARY INTRODUCTION TO STOCHASTIC INTEREST RATE MODELING RATE MODELING 2nd Edition

ADVANCED SERIES ON STATISTICAL SCIENCE & APPLIED PROBABILITY

Editor: Ole E. Barndorff-Nielsen

Published

Vol. 4 Principles of Statistical Inference from a Neo-Fisherian Perspective
by L. Pace and A. Salvan

Vol. 5 Local Stereology
by Eva B. Vedel Jensen

Vol. 6 Elementary Stochastic Calculus — With Finance in View
by T. Mikosch

Vol. 7 Stochastic Methods in Hydrology: Rain, Landforms and Floods
eds. O. E. Barndorff-Nielsen et al.

Vol. 8 Statistical Experiments and Decisions: Asymptotic Theory
by A. N. Shiryaev and V. G. Spokoiny

Vol. 9 Non-Gaussian Merton–Black–Scholes Theory
by S. I. Boyarchenko and S. Z. Levendorskiĭ

Vol. 10 Limit Theorems for Associated Random Fields and Related Systems
by A. Bulinski and A. Shashkin

Vol. 11 Stochastic Modeling of Electricity and Related Markets
by F. E. Benth, J. ⬚altytè Benth and S. Koekebakker

Vol. 12 An Elementary Introduction to Stochastic Interest Rate Modeling
by N. Privault

Vol. 13 Change of Time and Change of Measure
by O. E. Barndorff-Nielsen and A. Shiryaev

Vol. 14 Ruin Probabilities (2nd Edition)
by S. Asmussen and H. Albrecher

Vol. 15 Hedging Derivatives
by T. Rheinländer and J. Sexton

Vol. 16 An Elementary Introduction to Stochastic Interest Rate Modeling (2nd Edition)
by N. Privault

Advanced Series on

Statistical Science &

Applied Probability

Vol. 16

AN ELEMENTARY INTRODUCTION TO STOCHASTIC INTEREST RATE MODELING 2nd Edition

Nicolas Privault

Nanyang Technological University, Singapore

World Scientific

NEW JERSEY · LONDON · SINGAPORE · BEIJING · SHANGHAI · HONG KONG · TAIPEI · CHENNAI

Published by

World Scientific Publishing Co. Pte. Ltd.

5 Toh Tuck Link, Singapore 596224

USA office: 27 Warren Street, Suite 401-402, Hackensack, NJ 07601

UK office: 57 Shelton Street, Covent Garden, London WC2H 9HE

British Library Cataloguing-in-Publication Data
A catalogue record for this book is available from the British Library.

Advanced Series on Statistical Science and Applied Probability — Vol. 16
AN ELEMENTARY INTRODUCTION TO STOCHASTIC INTEREST
RATE MODELING
Second Edition

Copyright © 2012 by World Scientific Publishing Co. Pte. Ltd.

ISBN-13 978-981-4390-85-9
ISBN-10 981-4390-85-2

Printed in Singapore by World Scientific Printers.

To Zaidah Jailani

Preface

The first edition of this introduction to the stochastic modeling of interest rates and to the pricing of related derivatives has been published by World Scientific in 2008. These financial topics have been the object of intense research over the last two decades as well as in recent years, and remain increasingly important in financial engineering and risk management.

This second edition follows the publication in 2010 by Nankai University Press of the Chinese translation of the first edition. It incorporates a complete revision of the text as well as the addition of a new introductory chapter on credit risk, and additional exercises with their solutions. The book is aimed at the advanced undergraduate and beginning graduate levels, assuming that the reader has already received an introduction to the basics of probability and stochastic calculus. The interest rate models considered range from short rate to forward rate models such as the Heath-Jarrow-Morton (HJM) and Brace-Gatarek-Musiela (BGM) models, for which an introduction to calibration is given. The focus is placed on a step by step introduction of concepts and explicit calculations, in particular for the pricing of associated derivatives such as caps and swaptions.

Let us describe shortly what the main objectives of interest rate modeling are. It is common knowledge that according to the rules of continuous time compounding of interests, the value V_t at time $t > 0$ of a bank account earning interests at fixed rate $r > 0$ given by

$$V_t = V_0 e^{rt}, \qquad t \in \mathbb{R}_+,$$

can be reformulated in differential form as

$$\frac{dV_t}{V_t} = rdt.$$

The reality of the financial world is however more complex as it allows interest rates to become functions of time that can be subject to random changes, in which case the value of V_t becomes

$$V_t = V_0 \exp\left(\int_0^t r_s ds\right),$$

where $(r_s)_{s\in\mathbb{R}_+}$ is a time-dependent random process, called here a short term interest rate process. This type of interest rates, known as short rates, can be modeled in various ways using stochastic differential equations.

Short term interest rates models are still not sufficient to the needs of financial institutions, who often request the possibility to agree at a present time t for a loan to be delivered over a future period of time $[T, S]$ at a rate $r(t, T, S)$, $t \leq T \leq S$. This adds another level of sophistication to the modeling of interest rates, introducing the need for *forward interest rates processes* $r(t, T, S)$ now depending on three time indices. The *instantaneous* forward rates, defined as $T \mapsto F(t, T) := r(t, T, T)$, can be viewed at fixed time t as functions of one single variable T, the maturity date.

Forward rate processes $r(t, T, S)$ are of special interest from a functional analytic point of view because they can be reinterpreted as processes $t \mapsto r(t, \cdot, \cdot)$ taking values in a function space of two variables. Thus the modeling of forward rates makes a heavy use of stochastic processes taking values in (infinite-dimensional) function spaces, adding another level of technical difficulty in comparison with standard equity models.

Let us turn to the contents of this text. The first two chapters are devoted to reviews of stochastic calculus and classical Black-Scholes pricing for options on equities. Indeed, the Black-Scholes formula is a fundamental tool for the pricing interest rate derivatives, especially in the BGM model where it can be used as an approximation tool.

Next, after a rapid presentation of short term interest rate models in Chapter 3, we turn to the definition and pricing of zero-coupon bonds in Chapter 4. Zero-coupon bonds can be directly constructed from short term interest rate processes and they provide the basis for the construction of forward rate processes.

Forward rates, instantaneous rates, and their modeling using function spaces (such as the Nelson-Siegel and Svensson spaces) are considered in

Chapter 5. The stochastic Heath-Jarrow-Morton model for the modeling of forward rates is described in Chapter 6, along with the related absence of arbitrage condition.

The construction of forward measures and its consequences on the pricing of interest rate derivatives are given in Chapter 7, with application to the pricing of bond options. The problem of estimation and fitting of interest rate curves is considered in Chapter 8, where a solution to this problem is presented via the introduction of two-factor models.

This second edition comprises a new Chapter 9 on credit default, including in particular pricing models in which default of a bond can occur at a random time τ. We also consider the associated options (credit default swaps) that are designed as a protection against default, and we refer to [Bielecki and Rutkowski (2002)] for a complete account of credit risk modeling.

The last two chapters 10 and 11 are respectively devoted to LIBOR markets and to the Brace-Gatarek-Musiela (BGM) model, with an overview of calibration. For simplicity of exposition our approach is restricted to Brownian one-factor models, and we refer to [Björk (2004)], [Brigo and Mercurio (2006)], [James and Webber (2001)], [Carmona and Tehranchi (2006)], [Schoenmakers (2005)] for more complete presentation of the theory of interest rate modeling, including multifactor models.

The book is completed by two appendices, Appendix A on mathematical prerequisites, and Appendix B on further developments and perspectives in the field. Complete solutions to the exercises proposed in each chapter are provided at the end of the book. Most exercises are originals, while some exercises of Chapters 2 and 3 are classical or derived from [Kijima (2003)] and [Øksendal (2003)].

Finally it should be mentioned that this text grew from lecture notes on stochastic interest models given in the Master of Science in Mathematics for Finance and Actuarial Science (MSMFAS) at City University of Hong Kong, after the author started studying the topic in the MathFi project at INRIA Paris-Rocquencourt, France. The material in this second edition has also been developed through teaching in the Master of Science in Financial Engineering of the Nanyang Business School, at the Nanyang Technological University, Singapore. I thank both institutions for excellent working conditions and the possibility to facilitate these courses, and the MathFi project for encouragements to study interest rate models. Thanks are also due to the participating students as well as to Ming Gao (City University of Hong Kong), Xiao Wei (Central University of Finance and Economics, Beijing), and Ubbo Wieserma (University of Reading), for their suggestions and a careful reading of the first edition, which led to many corrections and improvements.

Nicolas Privault
2012

Contents

Preface   vii

1. A Review of Stochastic Calculus   1

    1.1   Brownian Motion   1
    1.2   Stochastic Integration   2
    1.3   Quadratic Variation 8
    1.4   Itô's Formula 10
    1.5   Exercises   12

2. A Review of Black-Scholes Pricing and Hedging   15

    2.1   Call and Put Options   15
    2.2   Market Model and Portfolio   17
    2.3   PDE Method   18
    2.4   The Girsanov Theorem   20
    2.5   Martingale Method 23
    2.6   Exercises   30

3. Short Term Interest Rate Models   33

    3.1   Mean-Reverting Models   33
    3.2   Constant Elasticity of Variance (CEV) Models   34
    3.3   Time-Dependent Models   35
    3.4   Exercises   35

4. Pricing of Zero-Coupon Bonds   39

    4.1   Definition and Basic Properties   39
    4.2   Absence of Arbitrage and the Markov Property   40

4.3   Absence of Arbitrage and the Martingale Property   42
4.4   PDE Solution: Probabilistic Method   44
4.5   PDE Solution: Analytical Method   46
4.6   Numerical Simulations   47
4.7   Exercises   50

5. Forward Rate Modeling   55

5.1   Forward Contracts   55
5.2   Instantaneous Forward Rate   58
5.3   Short Rates   60
5.4   Parametrization of Forward Rates   61
5.5   Curve Estimation   62
5.6   Exercises   63

6. The Heath-Jarrow-Morton (HJM) Model   65

6.1   Restatement of Objectives   65
6.2   Forward Vasicek Rates   67
6.3   Spot Forward Rate Dynamics   72
6.4   The HJM Condition   73
6.5   Markov Property of Short Rates   76
6.6   The Hull-White Model   78
6.7   Exercises   79

7. The Forward Measure and Derivative Pricing   81

7.1   Forward Measure   81
7.2   Dynamics under the Forward Measure   85
7.3   Derivative Pricing   88
7.4   Inverse Change of Measure   92
7.5   Exercises   93

8. Curve Fitting and a Two-Factor Model   97

8.1   Curve Fitting   97
8.2   Deterministic Shifts 100
8.3   The Correlation Problem 101
8.4   Two-Factor Model 104
8.5   Exercises 111

9. A Credit Default Model 115

    9.1 Survival Probabilities 115
    9.2 Stochastic Default 117
    9.3 Defaultable Bonds 119
    9.4 Credit Default Swaps 120
    9.5 Exercises 122

10. Pricing of Caps and Swaptions on the LIBOR 125

    10.1 Pricing of Caplets and Caps 125
    10.2 Forward Rate Measure and Tenor Structure 127
    10.3 Swaps and Swaptions 131
    10.4 The London InterBank Offered Rates (LIBOR) Model . . 133
    10.5 Swap Rates on the LIBOR Market 134
    10.6 Forward Swap Measures 137
    10.7 Swaption Pricing on the LIBOR Market 142
    10.8 Exercises 143

11. The Brace-Gatarek-Musiela (BGM) Model 149

    11.1 The BGM Model 149
    11.2 Cap Pricing 152
    11.3 Swaption Pricing 153
    11.4 Calibration of the BGM Model 157
    11.5 Exercises 160

12. Appendix A: Mathematical Tools 163

13. Appendix B: Some Recent Developments 171

14. Solutions to the Exercises 175

Bibliography 221

Index 225

Author Index 227

Chapter 1

A Review of Stochastic Calculus

We include a review of Brownian motion and stochastic integrals since they are a key tool to the modeling of interest rate processes. For simplicity, our presentation of the stochastic integral is restricted to square-integrable processes and we refer the reader to more advanced texts such as e.g. [Protter (2005)] for a comprehensive introduction.

1.1 Brownian Motion

Let $(\Omega, \mathcal{F}, \mathbb{P})$ be a probability space. The modeling of random assets in finance is mainly based on stochastic processes, which are families $(X_t)_{t \in I}$ of random variables indexed by a time interval I.

First of all we recall the definition of Brownian motion, which is a fundamental example of a stochastic process.

Definition 1.1. *The standard Brownian motion is a stochastic process* $(B_t)_{t \in \mathbb{R}_+}$ *such that*

1. $B_0 = 0$ *almost surely.*
2. *The sample paths* $t \longmapsto B_t$ *are (almost surely) continuous.*
3. *For any finite sequence of times* $t_0 < t_1 < \cdots < t_n$, *the increments*
$$B_{t_1} - B_{t_0}, B_{t_2} - B_{t_1}, \ldots, B_{t_n} - B_{t_{n-1}}$$
   are independent.
4. *For any times* $0 \leq s < t$, $B_t - B_s$ *is normally distributed with mean zero and variance* $t - s$.

For convenience we will sometimes interpret Brownian motion as a random walk over infinitesimal time intervals of length dt, with increments ΔB_t

over $[t, t + dt]$ given by

$$\Delta B_t = \pm \sqrt{dt} \qquad (1.1)$$

with equal probabilities $1/2$.

In the sequel we let $(\mathcal{F}_t)_{t \in \mathbb{R}_+}$ denote the filtration (i.e. an increasing family of sub σ-algebras of $\mathcal{F}$, see Appendix A) generated by $(B_t)_{t \in \mathbb{R}_+}$, i.e.:

$$\mathcal{F}_t = \sigma(B_s \ : \ 0 \le s \le t), \qquad t \in \mathbb{R}_+.$$

The n-dimensional Brownian motion can be constructed as

$$(B_t^1, \ldots, B_t^n)_{t \in \mathbb{R}_+}$$

where $(B_t^1)_{t \in \mathbb{R}_+}, \ldots, (B_t^n)_{t \in \mathbb{R}_+}$ are independent copies of $(B_t)_{t \in \mathbb{R}_+}$.

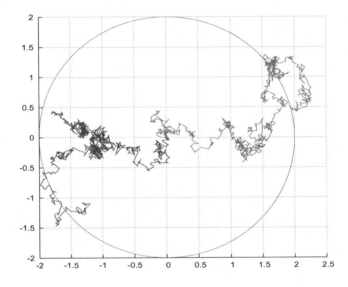

Fig. 1.1   Sample paths of a two-dimensional Brownian motion.

Next we turn to simulations of 2-dimensional, resp. 3-dimensional Brownian motion, cf. Figure 1.1, resp. Figure 1.2. Recall that the movement of pollen particles originally observed by R. Brown in 1827 was indeed 2-dimensional.

1.2   Stochastic Integration

In this section we construct the Itô stochastic integral of square-integrable adapted processes with respect to Brownian motion. The main use of stochastic integrals in finance is to model the behavior of a portfolio driven by a (random) risky asset.

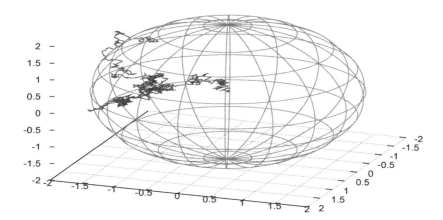

Fig. 1.2 Sample paths of a three-dimensional Brownian motion.

Definition 1.2. *A process $(X_t)_{t \in \mathbb{R}_+}$ is said to be $\mathcal{F}_t$-adapted if X_t is $\mathcal{F}_t$-measurable for all $t \in \mathbb{R}_+$.*

In other words, $(X_t)_{t \in \mathbb{R}_+}$ is $\mathcal{F}_t$-adapted when the value of X_t at time t only depends on information contained in the Brownian path up to time t.

Definition 1.3. *Let $L^p(\Omega \times \mathbb{R}_+)$ denote the space of p-integrable processes, i.e. the space of stochastic processes $u : \Omega \times \mathbb{R}_+ \to \mathbb{R}$ such that*

$$\mathbb{E}\left[\int_0^\infty |u_t|^p dt\right] < \infty,$$

and let $L^p_{ad}(\Omega \times \mathbb{R}_+)$, $p \in [1, \infty]$, denote the space of $\mathcal{F}_t$-adapted processes in $L^p(\Omega \times \mathbb{R}_+)$.

A naive definition of the stochastic integral with respect to Brownian motion would consist in writing

$$\int_0^\infty f(t) dB_t = \int_0^\infty f(t) \frac{dB_t}{dt} dt,$$

however this definition fails because the paths of Brownian motion are not differentiable:

$$\frac{dB_t}{dt} = \frac{\pm\sqrt{dt}}{dt} = \pm\frac{1}{\sqrt{dt}} \simeq \pm\infty.$$

Instead, stochastic integrals will be first constructed as integrals of simple predictable processes.

Definition 1.4. *Let $\mathcal{P}$ denote the space of simple predictable processes $(u_t)_{t \in \mathbb{R}_+}$ of the form*

$$u_t = \sum_{i=1}^{n} F_i \mathbf{1}_{(t_{i-1}^n, t_i^n]}(t), \qquad t \in \mathbb{R}_+, \tag{1.2}$$

where $F_i \in L^2(\Omega, \mathcal{F}_{t_{i-1}^n}, \mathbb{P})$ is $\mathcal{F}_{t_{i-1}^n}$-measurable, $i = 1, \ldots, n$.

One easily checks that the set $\mathcal{P}$ of simple predictable processes forms a linear space. From Lemma 1.1 of [Ikeda and Watanabe (1989)], p. 22 and p. 46, the space $\mathcal{P}$ of simple predictable processes is dense in $L_{ad}^p(\Omega \times \mathbb{R}_+)$ for any $p \geq 1$.

Proposition 1.1. *The stochastic integral with respect to Brownian motion $(B_t)_{t \in \mathbb{R}_+}$, defined on simple predictable processes $(u_t)_{t \in \mathbb{R}_+}$ of the form (1.2) by*

$$\int_0^\infty u_t dB_t := \sum_{i=1}^{n} F_i(B_{t_i} - B_{t_{i-1}}), \tag{1.3}$$

extends to $u \in L_{ad}^2(\Omega \times \mathbb{R}_+)$ via the isometry formula

$$\mathbb{E}\left[\int_0^\infty u_t dB_t \int_0^\infty v_t dB_t \right] = \mathbb{E}\left[\int_0^\infty u_t v_t dt \right], \tag{1.4}$$

$u, v \in L_{ad}^2(\Omega \times \mathbb{R}_+)$.

Proof.   We start by showing that the isometry (1.4) holds for the simple predictable process $u = \sum_{i=1}^n G_i \mathbf{1}_{(t_{i-1}, t_i]}$, with $0 = t_0 < t_1 < \cdots t_n$:

$$\mathbb{E}\left[\left(\int_0^\infty u_t dB_t \right)^2 \right] = \mathbb{E}\left[\left(\sum_{i=1}^n G_i(B_{t_i} - B_{t_{i-1}}) \right)^2 \right]$$

$$= \mathbb{E}\left[\sum_{i=1}^n |G_i|^2 (B_{t_i} - B_{t_{i-1}})^2 \right]$$

$$+ 2\mathbb{E}\left[\sum_{1 \leq i < j \leq n} G_i G_j (B_{t_i} - B_{t_{i-1}})(B_{t_j} - B_{t_{j-1}}) \right]$$

$$= \sum_{i=1}^n \mathbb{E}[\mathbb{E}[|G_i|^2 (B_{t_i} - B_{t_{i-1}})^2 | \mathcal{F}_{t_{i-1}}]]$$

$$+2 \sum_{1 \leq i < j \leq n} \mathbb{E}[\mathbb{E}[G_i G_j (B_{t_i} - B_{t_{i-1}})(B_{t_j} - B_{t_{j-1}})|\mathcal{F}_{t_{j-1}}]]$$

$$= \sum_{i=1}^{n} \mathbb{E}[|G_i|^2 \, \mathbb{E}[(B_{t_i} - B_{t_{i-1}})^2 |\mathcal{F}_{t_{i-1}}]]$$

$$+2 \sum_{1 \leq i < j \leq n} \mathbb{E}[G_i G_j (B_{t_i} - B_{t_{i-1}}) \, \mathbb{E}[(B_{t_j} - B_{t_{j-1}})|\mathcal{F}_{t_{j-1}}]]$$

$$= \mathbb{E}\left[\sum_{i=1}^{n} |G_i|^2 (t_i - t_{i-1})\right]$$

$$= \mathbb{E}[\|u\|_{L^2(\mathbb{R}_+)}^2].$$

The stochastic integral operator extends to $L^2_{ad}(\Omega \times \mathbb{R}_+)$ by density and a Cauchy sequence argument, applying the isometry (1.4). $\square$

In particular, (1.4) reads

$$\mathbb{E}\left[\left(\int_0^\infty u_t dB_t\right)^2\right] = \mathbb{E}\left[\int_0^\infty |u_t|^2 dt\right]$$

when $u = v$. In practice the stochastic integral (1.3) will be used to express the value of a self-financing portfolio as in Relation (2.10) below.

The Itô integral over the interval $[a, b]$ is defined as

$$\int_a^b u_s dB_s := \int_0^\infty \mathbf{1}_{[a,b]}(s) u_s dB_s, \qquad 0 \leq a \leq b,$$

for all $u \in L^2_{ad}(\Omega \times \mathbb{R}_+)$, with the relations

$$\int_a^c u_s dB_s = \int_a^b u_s dB_s + \int_b^c u_s dB_s, \qquad 0 \leq a \leq b \leq c,$$

and

$$\int_a^b dB_s = B_b - B_a, \qquad 0 \leq a \leq b.$$

Moreover the stochastic integral is a linear operator, i.e.:

$$\int_0^\infty (u_s + v_s) dB_s = \int_0^\infty u_s dB_s + \int_0^\infty v_s dB_s, \qquad u, v \in L^2_{ad}(\Omega \times \mathbb{R}_+).$$

The next proposition shows how to compute the conditional expectation of a stochastic integral by truncation of the integration interval.

Proposition 1.2. *For any $u \in L^2_{ad}(\Omega \times \mathbb{R}_+)$ we have*

$$\mathbb{E}\left[\int_0^\infty u_s dB_s \Big| \mathcal{F}_t\right] = \int_0^t u_s dB_s, \qquad t \in \mathbb{R}_+.$$

In particular, $\int_0^t u_s dB_s$ is $\mathcal{F}_t$-measurable, $t \in \mathbb{R}_+$.

Proof.   Let $u \in \mathcal{P}$ have the form $u = G\mathbf{1}_{(a,b]}$, where G is bounded and $\mathcal{F}_a$-measurable.

i) If $0 \leq a \leq t$ we have

$$
\begin{aligned}
\mathbb{E}\left[\int_0^\infty u_s dB_s \Big| \mathcal{F}_t\right] &= \mathbb{E}\left[G(B_b - B_a)|\mathcal{F}_t\right] \\
&= G\,\mathbb{E}\left[(B_b - B_a)|\mathcal{F}_t\right] \\
&= G\,\mathbb{E}\left[(B_b - B_t)|\mathcal{F}_t\right] + G\,\mathbb{E}\left[(B_t - B_a)|\mathcal{F}_t\right] \\
&= G(B_t - B_a) \\
&= \int_0^\infty \mathbf{1}_{[0,t]}(s)u_s dB_s.
\end{aligned}
$$

ii) If $0 \leq t \leq a$ we have for all bounded $\mathcal{F}_t$-measurable random variable F:

$$
\mathbb{E}\left[F\int_0^\infty u_s dB_s\right] = \mathbb{E}\left[FG(B_b - B_a)\right] = 0,
$$

hence

$$
\begin{aligned}
\mathbb{E}\left[\int_0^\infty u_s dB_s \Big| \mathcal{F}_t\right] &= \mathbb{E}\left[G(B_b - B_a)|\mathcal{F}_t\right] \\
&= 0 \\
&= \int_0^\infty \mathbf{1}_{[0,t]}(s)u_s dB_s.
\end{aligned}
$$

This statement is extended by linearity and density, since from the continuity of the conditional expectation on L^2 we have:

$$
\begin{aligned}
&\mathbb{E}\left[\left(\int_0^t u_s dB_s - \mathbb{E}\left[\int_0^\infty u_s dB_s \Big| \mathcal{F}_t\right]\right)^2\right] \\
&= \lim_{n\to\infty}\mathbb{E}\left[\left(\int_0^t u_s^n dB_s - \mathbb{E}\left[\int_0^\infty u_s dB_s \Big| \mathcal{F}_t\right]\right)^2\right] \\
&= \lim_{n\to\infty}\mathbb{E}\left[\left(\mathbb{E}\left[\int_0^\infty u_s^n dB_s - \int_0^\infty u_s dB_s \Big| \mathcal{F}_t\right]\right)^2\right] \\
&\leq \lim_{n\to\infty}\mathbb{E}\left[\mathbb{E}\left[\left(\int_0^\infty u_s^n dB_s - \int_0^\infty u_s dB_s\right)^2 \Big| \mathcal{F}_t\right]\right] \\
&\leq \lim_{n\to\infty}\mathbb{E}\left[\left(\int_0^\infty (u_s^n - u_s)dB_s\right)^2\right] \\
&= \lim_{n\to\infty}\mathbb{E}\left[\int_0^\infty |u_s^n - u_s|^2 ds\right] \\
&= 0.
\end{aligned}
$$

$\square$

In particular, since $\mathcal{F}_0 = \{\emptyset, \Omega\}$, the Itô integral is a centered random variable:

$$\mathbb{E}\left[\int_0^\infty u_s dB_s\right] = 0. \tag{1.5}$$

The following is an immediate corollary of Proposition 1.2.

Corollary 1.1. *The indefinite stochastic integral* $\left(\int_0^t u_s dB_s\right)_{t \in \mathbb{R}_+}$ *of* $u \in L^2_{ad}(\Omega \times \mathbb{R}_+)$ *is a martingale, i.e.:*

$$\mathbb{E}\left[\int_0^t u_\tau dB_\tau \Big| \mathcal{F}_s\right] = \int_0^s u_\tau dB_\tau, \quad 0 \leq s \leq t.$$

As an immediate consequence of the above corollary we have

$$\mathbb{E}\left[\int_t^\infty u_\tau dB_\tau \Big| \mathcal{F}_t\right] = 0, \quad \text{and} \quad \mathbb{E}\left[\int_0^t u_\tau dB_\tau \Big| \mathcal{F}_t\right] = \int_0^t u_\tau dB_\tau. \tag{1.6}$$

In particular, $\int_0^t u_\tau dB_\tau$ is $\mathcal{F}_t$-measurable for all $u \in L^2_{ad}(\Omega \times \mathbb{R}_+)$.

We close this section with a remark on the gaussianity of stochastic integrals of deterministic functions.

Proposition 1.3. *Let* $f \in L^2(\mathbb{R}_+)$. *The stochastic integral*

$$\int_0^\infty f(t) dB_t$$

is a Gaussian random variable with mean 0 *and variance*

$$\int_0^\infty |f(t)|^2 dt.$$

Proof. From the relation

$$\text{Var}(\alpha X) = \alpha^2 \text{Var}(X),$$

cf. (12.1) in Appendix A, the stochastic integral

$$\int_0^\infty f(t) dB_t := \sum_{k=1}^n a_k (B_{t_k} - B_{t_{k-1}}),$$

of the simple function

$$f(t) = \sum_{k=1}^n a_k \mathbf{1}_{(t_k, t_{k-1}]}(t),$$

has a centered Gaussian distribution with variance

$$\text{Var}\left[\int_0^\infty f(t)dB_t\right] = \sum_{k=1}^n a_k \,\text{Var}[B_{t_k} - B_{t_{k-1}}]$$

$$= \sum_{k=1}^n |a_k|^2 (t_k - t_{k-1})$$

$$= \sum_{k=1}^n |a_k|^2 \int_{t_{k-1}}^{t_k} dt$$

$$= \int_0^\infty |f(t)|^2 dt.$$

The result is extended by density of simple functions in $L^2(\mathbb{R}_+)$.   □

In particular, if $f \in L^2(\mathbb{R}_+)$ the Itô isometry (1.4) reads

$$E\left[\left(\int_0^\infty f(t)dB_t\right)^2\right] = \int_0^\infty |f(t)|^2 dt.$$

1.3 Quadratic Variation

We now introduce the notion of quadratic variation of Brownian motion.

Definition 1.5. *The quadratic variation of $(B_t)_{t\in\mathbb{R}_+}$ is the process $([B,B]_t)_{t\in\mathbb{R}_+}$ defined as*

$$[B,B]_t = B_t^2 - 2\int_0^t B_s dB_s, \quad t \in \mathbb{R}_+. \qquad (1.7)$$

Let now

$$\pi^n = \{0 = t_0^n < t_1^n < \cdots < t_{n-1}^n < t_n^n = t\}$$

denote a family of subdivision of $[0,t]$, such that

$$|\pi^n| := \max_{i=1,\dots,n} |t_i^n - t_{i-1}^n|$$

converges to 0 as n goes to infinity.

Proposition 1.4. *We have*

$$[B,B]_t = \lim_{n\to\infty} \sum_{i=1}^n (B_{t_i^n} - B_{t_{i-1}^n})^2, \qquad t \geq 0,$$

where the limit exists in $L^2(\Omega)$ and is independent of the sequence $(\pi^n)_{n\in\mathbb{N}}$ of subdivisions chosen.

Proof. As an immediate consequence of the Definition 1.3 of the stochastic integral we have

$$B_s(B_t - B_s) = \int_s^t B_s dB_\tau, \qquad 0 \le s \le t,$$

hence

$$[B, B]_{t_i^n} - [B, B]_{t_{i-1}^n} = B_{t_i^n}^2 - B_{t_{i-1}^n}^2 - 2 \int_{t_{i-1}^n}^{t_i^n} B_s dB_s$$

$$= (B_{t_i^n} - B_{t_{i-1}^n})^2 + 2 \int_{t_{i-1}^n}^{t_i^n} (B_{t_{i-1}^n} - B_s) dB_s,$$

hence

$$\mathbb{E}\left[\left([B, B]_t - \sum_{i=1}^n (B_{t_i^n} - B_{t_{i-1}^n})^2\right)^2\right]$$

$$= \mathbb{E}\left[\left(\sum_{i=1}^n [B, B]_{t_i^n} - [B, B]_{t_{i-1}^n} - (B_{t_i^n} - B_{t_{i-1}^n})^2\right)^2\right]$$

$$= 4\,\mathbb{E}\left[\left(\sum_{i=1}^n \int_0^t \mathbf{1}_{(t_{i-1}^n, t_i^n]}(s)(B_s - B_{t_{i-1}^n}) dB_s\right)^2\right]$$

$$= 4\,\mathbb{E}\left[\sum_{i=1}^n \int_{t_{i-1}^n}^{t_i^n} (B_s - B_{t_{i-1}^n})^2 ds\right]$$

$$= 4\,\mathbb{E}\left[\sum_{i=1}^n \int_{t_{i-1}^n}^{t_i^n} (s - t_{i-1}^n)^2 ds\right]$$

$$\le 4t|\pi^n|.$$

$\square$

In view of the informal construction (1.1) of Brownian motion as a random walk, the next proposition can be simply interpreted by writing $(\Delta B_t)^2 = dt$.

Proposition 1.5. *The quadratic variation of Brownian motion* $(B_t)_{t \in \mathbb{R}_+}$ *is*

$$[B, B]_t = t, \qquad t \in \mathbb{R}_+.$$

Proof. (cf. e.g. [Protter (2005)], Theorem I-28). For every subdivision

$$\{0 = t_0^n < \cdots < t_n^n = t\}$$

we have, by independence of the increments of Brownian motion:

$$
\mathbb{E}\left[\left(t - \sum_{i=1}^{n}(B_{t_i^n} - B_{t_{i-1}^n})^2\right)^2\right]
$$

$$
= \mathbb{E}\left[\left(\sum_{i=1}^{n}(B_{t_i^n} - B_{t_{i-1}^n})^2 - (t_i^n - t_{i-1}^n)\right)^2\right]
$$

$$
= \sum_{i=1}^{n}(t_i^n - t_{i-1}^n)^2 \,\mathbb{E}\left[\left(\frac{(B_{t_i^n} - B_{t_{i-1}^n})^2}{t_i^n - t_{i-1}^n} - 1\right)^2\right]
$$

$$
= \mathbb{E}[(Z^2 - 1)^2]\sum_{i=0}^{n}(t_i^n - t_{i-1}^n)^2
$$

$$
\leq t|\pi^n|\,\mathbb{E}[(Z^2 - 1)^2],
$$

where Z is a standard Gaussian random variable.   □

1.4   Itô's Formula

Using the rule $(dB_t)^2 = (\pm\sqrt{dt})^2 = dt$, Taylor's formula reads informally

$$
df(B_t) = f'(B_t)dB_t + \frac{1}{2}f''(B_t)(dB_t)^2
$$

$$
= f'(B_t)dB_t + \frac{1}{2}f''(B_t)dt.
$$

The Itô formula provides a generalization of this identity to processes X_t of the form

$$
X_t = X_0 + \int_0^t u_s dB_s + \int_0^t v_s ds, \qquad t \in \mathbb{R}_+,
$$

where u_t, v_t are adapted and sufficiently integrable processes.

The Itô formula can be stated in integral form as

$$
f(t, X_t) = f(0, X_0) + \int_0^t \frac{\partial f}{\partial x}(s, X_s)u_s dB_s \tag{1.8}
$$

$$
+ \int_0^t \frac{\partial f}{\partial x}(s, X_s)v_s ds + \int_0^t \frac{\partial f}{\partial s}(s, X_s)ds + \frac{1}{2}\int_0^t \frac{\partial^2 f}{\partial x^2}(s, X_s)u_s^2 ds,
$$

for $f \in \mathcal{C}^{1,2}(\mathbb{R}_+ \times \mathbb{R})$, or in differential form as:

$$
df(t, X_t) = \frac{\partial f}{\partial x}(t, X_t)u_t dB_t
$$

$$+\frac{\partial f}{\partial x}(t, X_t)v_t dt + \frac{\partial f}{\partial t}(t, X_t)dt + \frac{1}{2}\frac{\partial^2 f}{\partial x^2}(t, X_t)u_t^2 dt.$$

For the d-dimensional Brownian motion $(B_t)_{t\in\mathbb{R}_+}$, the Itô formula reads

$$f(B_t) = f(B_0) + \int_0^t \langle \nabla f(B_s), dB_s \rangle_H + \frac{1}{2}\int_0^t \Delta f(B_s)ds,$$

for all $\mathcal{C}^2$ functions f, where ∇ and Δ are respectively the gradient and laplacian operators on $\mathbb{R}^d$. Consider now two Itô processes X_t and Y_t of

the form

$$X_t = X_0 + \int_0^t u_s dB_s^{(1)} + \int_0^t v_s ds, \qquad t > 0,$$

and

$$Y_t = Y_0 + \int_0^t \xi_s dB_s^{(2)} + \int_0^t \zeta_s ds, \qquad t > 0,$$

where u_t, v_t, ξ_t, ζ_t are adapted and sufficiently integrable processes, and $(B_t^{(1)})_{t\in\mathbb{R}_+}$, $(B_t^{(2)})_{t\in\mathbb{R}_+}$, are two Brownian motions with correlation $\rho \in [-1,1]$, i.e. their covariation is

$$dB_t^{(1)} \cdot dB_t^{(2)} = \rho dt.$$

The Itô formula in two variables reads

$$f(t, X_t, Y_t)$$
$$= f(0, X_0, Y_0) + \int_0^t u_s \frac{\partial f}{\partial x}(s, X_s, Y_s)dB_s^{(1)} + \int_0^t \xi_s \frac{\partial f}{\partial y}(s, X_s, Y_s)dB_s^{(2)}$$
$$+ \int_0^t \frac{\partial f}{\partial s}(s, X_s, Y_s)ds + \int_0^t v_s \frac{\partial f}{\partial x}(s, X_s, Y_s)ds + \int_0^t \zeta_s \frac{\partial f}{\partial y}(s, X_s, Y_s)ds$$
$$+ \frac{1}{2}\int_0^t u_s^2 \frac{\partial^2 f}{\partial x^2}(s, X_s, Y_s)ds + \frac{1}{2}\int_0^t \xi_s^2 \frac{\partial^2 f}{\partial y^2}(s, X_s, Y_s)ds$$
$$+ \rho \int_0^t u_s \xi_s \frac{\partial^2 f}{\partial x \partial y}(s, X_s, Y_s)ds.$$

We close this chapter by quoting a classical result on stochastic differential equations, cf. e.g. [Protter (2005)], Theorem V-7. Let

$$\sigma : \mathbb{R}_+ \times \mathbb{R}^n \to \mathbb{R}^n \otimes \mathbb{R}^d$$

where $\mathbb{R}^n \otimes \mathbb{R}^d$ denotes the space of $d \times n$ matrices, and

$$b : \mathbb{R}_+ \times \mathbb{R}^n \to \mathbb{R}^n$$

satisfy the global Lipschitz condition

$$\|\sigma(t,x) - \sigma(t,y)\|^2 + \|b(t,x) - b(t,y)\|^2 \leq K^2 \|x - y\|^2,$$

$t \in \mathbb{R}_+$, $x, y \in \mathbb{R}^n$. Then there exists a unique strong solution to the stochastic differential equation

$$X_t = X_0 + \int_0^t \sigma(s, X_s)dB_s + \int_0^t b(s, X_s)ds,$$

where $(B_t)_{t \in \mathbb{R}_+}$ is a d-dimensional Brownian motion.

1.5 Exercises

Exercise 1.1. Let $c > 0$. Using the definition of Brownian motion $(B_t)_{t \in \mathbb{R}_+}$, show that:

(1) $(B_{c+t} - B_c)_{t \in \mathbb{R}_+}$ is a Brownian motion.
(2) $(cB_{t/c^2})_{t \in \mathbb{R}_+}$ is a Brownian motion.

Exercise 1.2. Solve the stochastic differential equation

$$dS_t = \mu S_t dt + \sigma S_t dB_t$$

where $\mu, \sigma \in \mathbb{R}$.

Exercise 1.3. Solve the stochastic differential equation

$$dX_t = -\alpha X_t dt + \sigma dB_t, \quad X_0 = 1,$$

with $\alpha > 0$ and $\sigma > 0$. *Hint.* Look for a solution of the form

$$X_t = a(t)\left(X_0 + \int_0^t b(s)dB_s\right),$$

where $a(\cdot)$ and $b(\cdot)$ are deterministic functions.

Exercise 1.4. Solve the stochastic differential equation

$$dX_t = tX_t dt + e^{t^2/2}dB_t, \quad X_0 = x_0.$$

Hint. Look for a solution of the form

$$X_t = a(t)\left(X_0 + \int_0^t b(s)dB_s\right),$$

where $a(\cdot)$ and $b(\cdot)$ are deterministic functions.

Exercise 1.5. Solve the stochastic differential equation

$$dY_t = (2\mu Y_t + \sigma^2)dt + 2\sigma\sqrt{Y_t}dB_t,$$

where $\mu \in \mathbb{R}$ and $\sigma > 0$. *Hint.* Let $X_t = \sqrt{Y_t}$.

Exercise 1.6. Let $f \in L^2([0,T])$. Compute the conditional expectation

$$E\left[e^{\int_0^T f(s)dB_s}\,\Big|\,\mathcal{F}_t\right], \qquad 0 \le t \le T,$$

where $(\mathcal{F}_t)_{t\in[0,T]}$ denotes the filtration generated by $(B_t)_{t\in[0,T]}$.

Exercise 1.7. Compute the expectation

$$E\left[\exp\left(\beta\int_0^T B_t dB_t\right)\right]$$

for all $\beta < 1/T$. *Hint*: Expand $(B_T)^2$ using Itô's formula.

Exercise 1.8. Given $T > 0$, let $(X_t^T)_{t\in[0,T]}$ denote the solution of the stochastic differential equation

$$dX_t^T = \sigma dB_t - \frac{X_t^T}{T-t}dt, \qquad 0 \le t \le T,$$

under the initial condition $X_0^T = 0$ and $\sigma > 0$.

(1) Show that

$$X_t^T = \sigma(T-t)\int_0^t \frac{1}{T-s}dB_s, \qquad 0 \le t \le T. \qquad (1.9)$$

Hint: Start by computing $d(X_t^T/(T-t))$ using Itô's calculus.

(2) Show that $\mathbb{E}[X_t^T] = 0$ for all $t \in [0,T]$.

(3) Show that $\operatorname{Var}[X_t^T] = \sigma^2 t(T-t)/T$ for all $t \in [0,T]$.

(4) Show that $X_T^T = 0$. The process $(X_t^T)_{t\in[0,T]}$ is called a *Brownian bridge*.

Chapter 2

A Review of Black-Scholes Pricing and Hedging

The Black-Scholes formula can be considered as a building block for the pricing of financial derivatives, and its importance is not restricted to the pricing of options on stocks. Indeed, the complexity of the interest rate models makes it in general difficult to obtain closed form expressions, and in many situations one has to rely on the Black-Scholes framework in order to find pricing formulas for interest rate derivatives, in particular in the BGM model, cf. Chapter 10.

2.1 Call and Put Options

An important concern for the buyer of a stock at time t is whether its price S_T can fall down at some future date T. The buyer may seek protection from a market crash by buying a contract that allows him to sell his asset at time T at a guaranteed price K fixed at an initial time t.

This contract is called a put option with strike price K and exercise date T. In case the price S_T falls down below the level K, exercising the contract will give the buyer of the option a gain equal to $K - S_T$ in comparison to others who did not subscribe the option. In turn, the issuer of the option will register a loss also equal to $K - S_T$, assuming the absence of transaction costs and other fees.

In the general case, the payoff of a (so-called European) put option will be of the form

$$
(K - S_T)^+ = \begin{cases} K - S_T & \text{if } S_T \leq K, \\ 0 & \text{if } S_T \geq K. \end{cases}
$$

In order for this contract to be fair, the buyer of the option should pay a fee (similar to an insurance fee) at the signature of the contract. The computation of this fee is an important issue, which is known as option pricing.

Two possible scenarios, with S_T finishing above K or below K, are illustrated in Figure 2.1.

Fig. 2.1   Samples price processes simulated according to a geometric Brownian motion.

On the other hand, if the trader aims at buying some stock or commodity, his interest will be in prices not going up and he might want to purchase a call option, which is a contract allowing him to buy the considered asset at time T at a price not higher than a level K fixed at time t.

Here, in the event that S_T goes above K, the buyer of the option will register a potential gain equal to $S_T - K$ in comparison to an agent who did not subscribe to the call option.

In general, a (so-called European) call option is an option with payoff function

$$\phi(S_T) = (S_T - K)^+ = \begin{cases} S_T - K & \text{if } S_T \geq K, \\ \\ 0, & \text{if } S_T \leq K. \end{cases}$$

In connection with the interest rate models to be presented in the next chapters, we note at the present stage that similar contracts can be applied to interest rates.

A contract protecting a borrower at variable rate r_t by forcing his offered rate not to go above a level κ will result into an interest rate equal to $\min(r_t, \kappa)$. The corresponding contract is called an interest rate cap and potentially gives its buyer an advantage $(r_t - \kappa)^+$, measured in terms of interest rate points. The counterpart of a cap is called a floor and offers a similar protection, this time against interest rates going down, for the benefit of lenders.

The classical Black-Scholes formula is of importance for the pricing of interest rates derivatives since some of the interest rate models that we will consider will be based on geometric Brownian motion.

2.2 Market Model and Portfolio

Let $r : \mathbb{R}_+ \longrightarrow \mathbb{R}$, $\mu : \mathbb{R}_+ \longrightarrow \mathbb{R}$ and $\sigma : \mathbb{R}_+ \longrightarrow (0, \infty)$ be deterministic non negative bounded functions. Let $(A_t)_{t \in \mathbb{R}_+}$ be a riskless asset with price given by

$$\frac{dA_t}{A_t} = r_t dt, \qquad A_0 = 1, \qquad t \in \mathbb{R}_+, \qquad (2.1)$$

i.e.

$$A_t = A_0 \exp \left(\int_0^t r_s ds \right), \qquad t \in \mathbb{R}_+.$$

For $t > 0$, let $(S_t)_{t \in [0,T]}$ be the price process defined by the stochastic differential equation

$$dS_t = \mu_t S_t dt + \sigma_t S_t dB_t, \qquad t \in \mathbb{R}_+,$$

i.e. in integral form:

$$S_t = S_0 + \int_0^t \mu_u S_u du + \int_0^t \sigma_u S_u dB_u, \qquad t \in \mathbb{R}_+,$$

with solution

$$S_t = S_0 \exp\left(\int_0^t \sigma_u dB_u + \int_0^t \left(\mu_u - \frac{1}{2}\sigma_u^2 \right) du \right),$$

$t \in \mathbb{R}_+$, cf. Exercise 1.2.

Let η_t and ζ_t be the numbers of units invested at time t, respectively in the assets $(S_t)_{t \in \mathbb{R}_+}$ and $(A_t)_{t \in \mathbb{R}_+}$. The value of the portfolio V_t at time t is given by

$$V_t = \zeta_t A_t + \eta_t S_t, \qquad t \in \mathbb{R}_+. \tag{2.2}$$

Definition 2.1. *The portfolio V_t is said to be self-financing if*

$$dV_t = \zeta_t dA_t + \eta_t dS_t. \tag{2.3}$$

Note that the self-financing condition (2.3) can be written as

$$A_t d\zeta_t + S_t d\eta_t = 0, \qquad 0 \leq t \leq T$$

provided one neglects the bracket $d\langle S, \eta \rangle_t$.

2.3   PDE Method

In this standard Black-Scholes model it is possible to determine a portfolio strategy for the hedging of European claims. First, note that the self-financing condition (2.3) implies

$$\begin{aligned}
dV_t &= \zeta_t dA_t + \eta_t dS_t \\
&= r_t \zeta_t A_t dt + \mu_t \eta_t S_t dt + \sigma_t \eta_t S_t dB_t \\
&= r_t V_t dt + (\mu_t - r_t)\eta_t S_t dt + \sigma_t \eta_t S_t dB_t,
\end{aligned} \tag{2.4}$$

$t \in \mathbb{R}_+$. Assume now that the value V_t of the portfolio at time t is given by a function $C(t, x)$ as

$$V_t = C(t, S_t), \qquad t \in \mathbb{R}_+.$$

An application of the Itô formula (1.8) leads to

$$\begin{aligned}
dC(t, S_t) &= \left(\frac{\partial C}{\partial t} + \mu_t S_t \frac{\partial C}{\partial x} + \frac{1}{2}\frac{\partial^2 C}{\partial x^2}S_t^2 \sigma_t^2 \right)(t, S_t)dt \\
&\quad + \sigma_t S_t \frac{\partial C}{\partial x}(t, S_t)dB_t.
\end{aligned} \tag{2.5}$$

Therefore, after respective identification of the terms in dB_t and dt in (2.4) and (2.5) we get

$$
\begin{cases}
r_t C(t, S_t) = \left(\dfrac{\partial C}{\partial t} + r_t S_t \dfrac{\partial C}{\partial x} + \dfrac{1}{2} \sigma_t^2 S_t^2 \dfrac{\partial^2 C}{\partial x^2} \right)(t, S_t), \\[2ex]
\eta_t S_t \sigma_t dB_t = S_t \sigma_t \dfrac{\partial C}{\partial x}(t, S_t) dB_t,
\end{cases}
\tag{2.6}
$$

hence

$$
\eta_t = \frac{\partial C}{\partial x}(t, S_t).
\tag{2.7}
$$

The process $(\eta_t)_{t\in\mathbb{R}_+}$ is called the Delta. In addition to computing the Delta we derived the Black-Scholes partial differential equation (PDE), as stated in the next proposition.

Proposition 2.1. *The Black-Scholes PDE for the price of a European call is written as*

$$
\frac{\partial C}{\partial t}(t, x) + r_t x \frac{\partial C}{\partial x}(t, x) + \frac{1}{2} x^2 \sigma_t^2 \frac{\partial^2 C}{\partial x^2}(t, x) = r_t C(t, x),
$$

under the terminal condition $C(T, x) = (x - K)^+$.

The solution of this PDE is given by the Black-Scholes formula

$$
C(t, x) = \mathrm{Bl}(K, x, \tilde{\sigma}_t, \tilde{r}_t, T - t) := x\Phi(d_1) - Ke^{-(T-t)\tilde{r}_t}\Phi(d_2),
\tag{2.8}
$$

where

$$
\Phi(x) = \frac{1}{\sqrt{2\pi}} \int_{-\infty}^{x} e^{-y^2/2} dy, \qquad x \in \mathbb{R},
$$

denotes the Gaussian distribution function,

$$
d_1 = \frac{\log(x/K) + (\tilde{r}_t + \tilde{\sigma}_t^2/2)(T-t)}{\tilde{\sigma}_t \sqrt{T-t}}, \quad d_2 = \frac{\log(x/K) + (\tilde{r}_t - \tilde{\sigma}_t^2/2)(T-t)}{\tilde{\sigma}_t \sqrt{T-t}},
$$

and

$$
\tilde{\sigma}_t^2 = \frac{1}{T-t} \int_t^T |\sigma(s)|^2 ds, \qquad \tilde{r}_t = \frac{1}{T-t} \int_t^T r(s) ds.
$$

We refer to [Mikosch (1998)] and [Øksendal (2003)] for more detailed expositions of these topics.

2.4   The Girsanov Theorem

Before proceeding to the pricing of options using the martingale approach, we need to review the Girsanov theorem. Let us come back to the informal interpretation (1.1) of Brownian motion via its infinitesimal increments:

$$\Delta B_t = \pm\sqrt{dt},$$

with

$$\mathbb{P}(\Delta B_t = +\sqrt{dt}) = \mathbb{P}(\Delta B_t = -\sqrt{dt}) = \frac{1}{2}.$$

Fig. 2.2   Drifted Brownian path.

Clearly, given $\nu \in \mathbb{R}$, the drifted process $\nu t + B_t$ is no longer a standard Brownian motion because it is not centered:

$$\mathbb{E}[\nu t + B_t] = \nu t + \mathbb{E}[B_t] = \nu t \neq 0,$$

cf. Figure 2.2. This identity can be formulated in terms of infinitesimal increments as

$$\mathbb{E}[\nu dt + dB_t] = \frac{1}{2}(\nu dt + \sqrt{dt}) + \frac{1}{2}(\nu dt - \sqrt{dt}) = \nu dt \neq 0.$$

In order to make $\nu t + B_t$ a centered process (i.e. a standard Brownian motion, since $\nu t + B_t$ conserves all the other properties (1)-(3) in Definition 1.1, one may change the probabilities of ups and downs, which have been fixed so far equal to 1/2.

That is, the problem is now to find two numbers $p, q \in [0, 1]$ such that

$$\begin{cases} p(\nu dt + \sqrt{dt}) + q(\nu dt - \sqrt{dt}) = 0 \\ \\ p + q = 1. \end{cases}$$

The solution to this problem is given by

$$p = \frac{1}{2}(1 - \nu\sqrt{dt}) \quad \text{and} \quad q = \frac{1}{2}(1 + \nu\sqrt{dt}).$$

Still considering Brownian motion as a discrete random walk with independent increments $\pm\sqrt{dt}$, the corresponding probability density will be obtained by taking the product of the above probabilities divided by $1/2^N$, that is:

$$2^N \prod_{0 < t < T} \left(\frac{1}{2} \mp \frac{1}{2}\nu\sqrt{dt} \right)$$

where 2^N is a normalization factor and $N = T/dt$ is the (infinitely large) number of discrete time steps. Using elementary calculus, this density can be informally shown to converge as follows:

$$2^N \prod_{0 < t < T} \left(\frac{1}{2} \mp \frac{1}{2}\nu\sqrt{dt} \right) = \prod_{0 < t < T} \left(1 \mp \nu\sqrt{dt} \right)$$

$$= \exp\left(\log \prod_{0 < t < T} \left(1 \mp \nu\sqrt{dt} \right) \right)$$

$$= \exp\left(\sum_{0 < t < T} \log \left(1 \mp \nu\sqrt{dt} \right) \right)$$

$$\simeq \exp\left(\nu \sum_{0 < t < T} \mp\sqrt{dt} - \frac{1}{2} \sum_{0 < t < T} (\mp\nu\sqrt{dt})^2 \right)$$

$$= \exp\left(\nu \sum_{0 < t < T} \mp\sqrt{dt} - \frac{1}{2}\nu^2 \sum_{0 < t < T} dt \right)$$

$$= \exp\left(-\nu B_T - \frac{1}{2}\nu^2 T \right).$$

The Girsanov theorem can be restated as follows in a more rigorous way. Recall that here, $\Omega = \mathcal{C}_0([0, T])$ is the Wiener space and $\omega \in \Omega$ is a continuous function on $[0, T]$ starting at 0 in $t = 0$. Consider the probability $\mathbb{Q}$ defined by

$$d\mathbb{Q}(\omega) = \exp\left(-\nu B_T - \frac{1}{2}\nu^2 T \right) d\mathbb{P}(\omega).$$

Then the process $\nu t + B_t$ is a standard (centered) Brownian motion under $\mathbb{Q}$.

For example, the fact that $\nu T + B_T$ has a standard (centered) Gaussian law under $\mathbb{Q}$ can be recovered as follows:

$$
\begin{aligned}
\mathbb{E}_{\mathbb{Q}}[f(\nu T + B_T)] &= \int_{\Omega} f(\nu T + B_T)d\mathbb{Q} \\
&= \int_{\Omega} f(\nu T + B_T)\exp\left(-\nu B_T - \frac{1}{2}\nu^2 T\right)d\mathbb{P} \\
&= \int_{-\infty}^{\infty} f(\nu T + x)\exp\left(-\nu x - \frac{1}{2}\nu^2 T\right)e^{-\frac{x^2}{2T}}\frac{dx}{\sqrt{2\pi T}} \\
&= \int_{-\infty}^{\infty} f(y)e^{-\frac{y^2}{2T}}\frac{dy}{\sqrt{2\pi T}} \\
&= \int_{\Omega} f(B_T)d\mathbb{P} \\
&= \mathbb{E}_{\mathbb{P}}[f(B_T)].
\end{aligned}
$$

The above change of variable also admits the following informal infinite-dimensional formulation:

$$
\begin{aligned}
\prod_{0<t<T} \frac{e^{-\frac{1}{2dt}(dB_t+\psi_t dt)^2}}{\sqrt{2\pi dt}} &= \prod_{0<t<T} e^{-\frac{1}{2dt}(2\psi_t dB_t dt + \psi_t^2 dt^2)}\frac{e^{-\frac{1}{2dt}(dB_t)^2}}{\sqrt{2\pi dt}} \\
&= \prod_{0<t<T} e^{-\psi_t dB_t - \frac{1}{2}\psi_t^2 dt}\frac{e^{-\frac{1}{2dt}(dB_t)^2}}{\sqrt{2\pi dt}} \\
&= e^{-\int_0^T \psi_t dB_t - \frac{1}{2}\int_0^T \psi_t^2 dt}\prod_{0<t<T}\frac{e^{-\frac{1}{2dt}(dB_t)^2}}{\sqrt{2\pi dt}}.
\end{aligned}
$$

More precisely, the Girsanov theorem can actually be extended to shifts by adapted processes as follows, cf. e.g. [Protter (2005)], Theorem III-42.

Theorem 2.1. *Let $(\psi_t)_{t\in[0,T]}$ be an adapted process satisfying the Novikov integrability condition*

$$
\mathbb{E}\left[\exp\left(\frac{1}{2}\int_0^T |\psi_t|^2 dt\right)\right] < \infty, \tag{2.9}
$$

and let $\mathbb{Q}$ denote the probability measure defined by

$$
\frac{d\mathbb{Q}}{d\mathbb{P}} = \exp\left(-\int_0^T \psi_s dB_s - \frac{1}{2}\int_0^T \psi_s^2 ds\right).
$$

Then

$$
\hat{B}_t := B_t + \int_0^t \psi_s ds, \qquad 0 \le t \le T,
$$

is a standard Brownian motion under $\mathbb{Q}$.

2.5 Martingale Method

In this section we give the expression of the Black-Scholes price using expectations of discounted payoffs.

Definition 2.2. *A market is said without arbitrage if there exists (at least) a probability $\mathbb{Q}$ under which the discounted price process*

$$\tilde{S}_t := \exp\left(-\int_0^t r_s ds\right) S_t, \qquad t \in \mathbb{R}_+,$$

is a martingale under $\mathbb{Q}$.

Such a probability $\mathbb{Q}$ is usually called a risk-neutral probability or a martingale measure. When the martingale measure is unique, the market is said to be *complete*. We will now show that the Black-Scholes model admits a unique martingale measure, which shows that the market is without arbitrage and complete.

Let now $(\psi_t)_{t \in [0,T]}$ be defined as

$$\psi_t := \frac{\mu_t - r_t}{\sigma_t}, \qquad 0 \leq t \leq T,$$

and let $\mathbb{Q}$ denote the probability defined by

$$\frac{d\mathbb{Q}}{d\mathbb{P}} = \exp\left(-\int_0^T \psi_s dB_s - \frac{1}{2}\int_0^T \psi_s^2 ds\right).$$

From the Girsanov Theorem 2.1 we know that

$$\hat{B}_t := B_t + \int_0^t \psi_s ds, \qquad 0 \leq t \leq T,$$

is a Brownian motion under $\mathbb{Q}$. Let also

$$\tilde{V}_t = V_t \exp\left(-\int_0^t r_s ds\right), \quad \text{and} \quad \tilde{S}_t = S_t \exp\left(-\int_0^t r_s ds\right),$$

denote the discounted portfolio and underlying asset.

Lemma 2.1. *The following statements are equivalent:*

i) the portfolio V_t is self-financing,

ii) we have

$$\tilde{V}_t = \tilde{V}_0 + \int_0^t \sigma_u \eta_u \tilde{S}_u d\hat{B}_u, \qquad t \in \mathbb{R}_+, \tag{2.10}$$

iii) we have

$$V_t = V_0 \exp\left(\int_0^t r_u du\right) + \int_0^t \sigma_u \eta_u S_u \exp\left(\int_u^t r_s ds\right) d\hat{B}_u, \qquad t \in \mathbb{R}_+.$$
$$(2.11)$$

Proof. First, note that (2.10) is clearly equivalent to (2.11). Now, the self-financing condition (2.3) shows that

$$\begin{aligned} dV_t &= \zeta_t dA_t + \eta_t dS_t \\ &= \zeta_t A_t r_t dt + \eta_t r_t S_t dt + \sigma_t \eta_t S_t d\hat{B}_t \\ &= r_t V_t dt + \sigma_t \eta_t S_t d\hat{B}_t, \qquad t \in \mathbb{R}_+, \end{aligned}$$

hence

$$\begin{aligned} d\tilde{V}_t &= d\left(\exp\left(-\int_0^t r_s ds\right) V_t\right) \\ &= -r_t \exp\left(-\int_0^t r_s ds\right) V_t dt + \exp\left(-\int_0^t r_s ds\right) dV_t \\ &= \exp\left(-\int_0^t r_s ds\right) \sigma_t \eta_t S_t d\hat{B}_t, \qquad t \in \mathbb{R}_+, \end{aligned}$$

i.e. (2.10) holds. Conversely, if (2.10) is satisfied we have

$$\begin{aligned} dV_t &= d\left(\exp\left(\int_0^t r_s ds\right) \tilde{V}_t\right) \\ &= r_t \exp\left(\int_0^t r_s ds\right) \tilde{V}_t dt + \exp\left(\int_0^t r_s ds\right) d\tilde{V}_t \\ &= r_t \exp\left(\int_0^t r_s ds\right) \tilde{V}_t dt + \sigma_t \eta_t S_t d\hat{B}_t \\ &= V_t r_t dt + \sigma_t \eta_t S_t d\hat{B}_t \\ &= \zeta_t A_t r_t dt + \eta_t S_t r_t dt + \sigma_t \eta_t S_t d\hat{B}_t \\ &= \zeta_t dA_t + \eta_t dS_t, \end{aligned}$$

hence the portfolio is self-financing.   □

In the next proposition we compute a self-financing hedging strategy leading to an arbitrary square-integrable random variable F admitting a predictable representation of the form

$$F = \mathbb{E}_\mathbb{Q}[F] + \int_0^T \xi_t d\hat{B}_t,$$
$$(2.12)$$

where $(\xi_t)_{t \in [0,t]}$ is a square-integrable adapted process.

Proposition 2.2. *Given $F \in L^2(\Omega)$, let*

$$\eta_t = \frac{\exp\left(-\int_t^T r_s ds\right)}{\sigma_t S_t} \xi_t, \tag{2.13}$$

$$\zeta_t = \frac{\exp\left(-\int_t^T r_u du\right) \mathbb{E}_{\mathbb{Q}}[F|\mathcal{F}_t] - \eta_t S_t}{A_t}, \qquad 0 \le t \le T. \tag{2.14}$$

Then the portfolio $(\eta_t, \zeta_t)_{t \in [0,T]}$ is self-financing, and letting

$$V_t = \zeta_t A_t + \eta_t S_t, \qquad 0 \le t \le T, \tag{2.15}$$

we have

$$V_t = \exp\left(-\int_t^T r_u du\right) \mathbb{E}_{\mathbb{Q}}[F|\mathcal{F}_t], \qquad 0 \le t \le T. \tag{2.16}$$

In particular we have

$$V_T = F,$$

i.e. the portfolio yields a hedging strategy leading to F, starting from the initial value

$$V_0 = \exp\left(-\int_0^T r_u du\right) \mathbb{E}_{\mathbb{Q}}[F].$$

Proof. Applying (2.14) and (2.15) at $t = 0$ we get

$$\mathbb{E}_{\mathbb{Q}}[F] \exp\left(-\int_0^T r_u du\right) = \zeta_0 A_0 + \eta_0 S_0 = V_0,$$

hence from (2.14) again, the definition (2.13) of η_t and (2.12), we obtain

$$V_t = \zeta_t A_t + \eta_t S_t$$

$$= \exp\left(-\int_t^T r_u du\right) \mathbb{E}_{\mathbb{Q}}[F|\mathcal{F}_t]$$

$$= \exp\left(-\int_t^T r_u du\right) \left(\mathbb{E}_{\mathbb{Q}}[F] + \int_0^t \xi_u d\hat{B}_u\right)$$

$$= V_0 \exp\left(\int_0^t r_u du\right) + \exp\left(-\int_t^T r_u du\right) \int_0^t \xi_u d\hat{B}_u$$

$$= V_0 \exp\left(\int_0^t r_u du\right) + \int_0^t \eta_u \sigma_u S_u \exp\left(\int_u^t r_s ds\right) d\hat{B}_u, \qquad 0 \le t \le T,$$

and from Lemma 2.1 this also implies that the portfolio $(\eta_t, \zeta_t)_{t \in [0,T]}$ is self-financing. □

The above proposition shows that there always exists a hedging strategy starting from

$$V_0 = \mathbb{E}_{\mathbb{Q}}[F] \exp\left(-\int_0^T r_u du\right).$$

In addition, since there exists a hedging strategy leading to

$$\tilde{V}_T = F \exp\left(-\int_0^T r_u du\right),$$

then by (2.10), $(\tilde{V}_t)_{t \in [0,T]}$ is necessarily a martingale with

$$\tilde{V}_t = \mathbb{E}_{\mathbb{Q}}[\tilde{V}_T | \mathcal{F}_t] = \exp\left(-\int_0^T r_u du\right) \mathbb{E}_{\mathbb{Q}}[F|\mathcal{F}_t], \qquad 0 \leq t \leq T,$$

and initial value

$$\tilde{V}_0 = \mathbb{E}_{\mathbb{Q}}[\tilde{V}_T] = \mathbb{E}_{\mathbb{Q}}[F] \exp\left(-\int_0^T r_u du\right).$$

In practice, the hedging problem can now be reduced to the computation of the process $(\xi_t)_{t \in [0,T]}$ appearing in (2.12). This computation, called the Delta hedging, can be performed by application of the Itô formula and the Markov property, see e.g. [Protter (2001)]. Consider the (non homogeneous) semi-group $(P_{s,t})_{0 \leq s \leq t \leq T}$ associated to $(S_t)_{t \in [0,T]}$ and defined by

$$P_{s,t}f(S_s) = \mathbb{E}_{\mathbb{Q}}[f(S_t) \mid S_s]$$
$$= \mathbb{E}_{\mathbb{Q}}[f(S_t) \mid \mathcal{F}_s], \quad 0 \leq s \leq t \leq T,$$

which acts on $\mathcal{C}_b^2(\mathbb{R})$ functions, with

$$P_{s,t}P_{t,u} = P_{s,u}, \qquad 0 \leq s \leq t \leq u \leq T.$$

Note that $(P_{t,T}f(S_t))_{t \in [0,T]}$ is an $\mathcal{F}_t$-martingale, i.e.:

$$\mathbb{E}_{\mathbb{Q}}[P_{t,T}f(S_t) \mid \mathcal{F}_s] = \mathbb{E}_{\mathbb{Q}}[\mathbb{E}_{\mathbb{Q}}[f(S_T) \mid \mathcal{F}_t] \mid \mathcal{F}_s]$$
$$= \mathbb{E}_{\mathbb{Q}}[f(S_T) \mid \mathcal{F}_s]$$
$$= P_{s,T}f(S_s), \qquad (2.17)$$

$0 \leq s \leq t \leq T$. The next lemma allows us to compute the process $(\xi_t)_{t \in [0,T]}$ in case the payoff F is of the form $F = \phi(S_T)$ for some function ϕ.

Lemma 2.2. *Let* $\phi \in \mathcal{C}_b^2(\mathbb{R})$. *The predictable representation*

$$\phi(S_T) = \mathbb{E}_{\mathbb{Q}}[\phi(S_T)] + \int_0^T \xi_t d\hat{B}_t \qquad (2.18)$$

is given by

$$\xi_t = \sigma_t S_t \frac{\partial}{\partial x}(P_{t,T}\phi)(S_t), \quad 0 \leq t \leq T. \qquad (2.19)$$

Proof. Since $P_{t,T}\phi$ is in $\mathcal{C}^2(\mathbb{R})$, we can apply the Itô formula (1.8) to the process

$$t \longmapsto P_{t,T}\phi(S_t) = \mathbb{E}_{\mathbb{Q}}[\phi(S_T) \mid \mathcal{F}_t], \qquad (2.20)$$

which is a martingale from (2.17), cf. also Appendix A. From the fact that the finite variation term in the Itô formula vanishes when $(P_{t,T}\phi(S_t))_{t\in[0,T]}$ is a martingale, (see e.g. Corollary 1, p. 72 of [Protter (2005)]), we obtain:

$$P_{t,T}\phi(S_t) = P_{0,T}\phi(S_0) + \int_0^t \sigma_s S_s \frac{\partial}{\partial x}(P_{s,T}\phi)(S_s)d\hat{B}_s, \quad 0 \le t \le T, \ (2.21)$$

with $P_{0,T}\phi(S_0) = \mathbb{E}_{\mathbb{Q}}[\phi(S_T)]$. Letting $t = T$, we obtain (2.19) by uniqueness of the predictable representation (2.18) of $F = \phi(S_T)$. $\qquad\square$

Let now $(S_{t,s}^x)_{s\in[t,\infty)}$ be the price process solution of the stochastic differential equation

$$\frac{dS_{t,s}^x}{S_{t,s}^x} = r_s ds + \sigma_s d\hat{B}_s, \qquad s \in [t, \infty),$$

with initial condition $S_{t,t}^x = x \in (0, \infty)$.

The value V_t of the portfolio at time $t \in [0,T]$ can be computed from (2.16) as

$$V_t = \exp\left(-\int_t^T r_u du\right) \mathbb{E}_{\mathbb{Q}}[\phi(S_T)|\mathcal{F}_t]$$
$$= C(t, S_t),$$

where

$$C(t,x) = e^{-(T-t)\bar{r}_t}\, \mathbb{E}_{\mathbb{Q}}[\phi(S_T)|S_t = x]$$
$$= e^{-(T-t)\bar{r}_t}\, P_{t,T}\phi(x)$$
$$= e^{-(T-t)\bar{r}_t}\, \mathbb{E}_{\mathbb{Q}}[\phi(S_{t,T}^x)],$$

$0 \le t \le T$, from Relation (12.5) in Appendix A. Again, from the fact that the finite variation term vanishes in (2.21) we recover the fact that $C(t,x)$ solves the Black-Scholes PDE:

$$\begin{cases} \dfrac{\partial C}{\partial t}(t,x) + \dfrac{1}{2}x^2\sigma^2(t)\dfrac{\partial^2 C}{\partial x^2}(t,x) + xr(t)\dfrac{\partial C}{\partial x}(t,x) = r(t)C(t,x), \\[2ex] C(T,x) = \phi(x). \end{cases}$$

In the case of European options with payoff function $\phi(x) = (x - K)^+$ we recover Relation (2.8), i.e.

$$C(t, x) = \mathrm{Bl}(K, x, \tilde{\sigma}_t, \tilde{r}_t, T - t),$$

as a consequence of (2.16) and the following lemma.

Lemma 2.3. *Let X be a centered Gaussian random variable with variance v^2. We have*

$$\mathbb{E}_{\mathbb{Q}}[(e^{m+X} - K)^+] = e^{m+\frac{v^2}{2}} \Phi(v + (m - \log K)/v) - K\Phi((m - \log K)/v).$$

Proof. We have

$$\mathbb{E}_{\mathbb{Q}}[(e^{m+X} - K)^+] = \int_{-\infty}^{\infty} (e^{m+x} - K)^+ e^{-\frac{x^2}{2v^2}} \frac{dx}{\sqrt{2\pi v^2}}$$

$$= \int_{-m+\log K}^{\infty} (e^{m+x} - K) e^{-\frac{x^2}{2v^2}} \frac{dx}{\sqrt{2\pi v^2}}$$

$$= e^m \int_{-m+\log K}^{\infty} e^{x - \frac{x^2}{2v^2}} \frac{dx}{\sqrt{2\pi v^2}} - K \int_{-m+\log K}^{\infty} e^{-\frac{x^2}{2v^2}} \frac{dx}{\sqrt{2\pi v^2}}$$

$$= e^{m+\frac{v^2}{2}} \int_{-m+\log K}^{\infty} e^{-\frac{(v^2-x)^2}{2v^2}} \frac{dx}{\sqrt{2\pi v^2}} - K \int_{(-m+\log K)/v}^{\infty} e^{-x^2/2} \frac{dx}{\sqrt{2\pi}}$$

$$= e^{m+\frac{v^2}{2}} \int_{-v^2-m+\log K}^{\infty} e^{-\frac{x^2}{2v^2}} \frac{dx}{\sqrt{2\pi v^2}} - K\Phi((m - \log K)/v)$$

$$= e^{m+\frac{v^2}{2}} \Phi(v + (m - \log K)/v) - K\Phi((m - \log K)/v). \qquad \square$$

Moreover, still in the case of European options, the process $(\xi)_{t \in [0,T]}$ can be computed via the next proposition.

Proposition 2.3. *Assume that $F = (S_T - K)^+$. Then for $0 \le t \le T$ we have*

$$\xi_t = \sigma_t \, \mathbb{E}_{\mathbb{Q}} \left[S_{t,T}^x \mathbf{1}_{[K,\infty[}(S_{t,T}^x) \right]_{x=S_t}.$$

Proof. This result follows from Lemma 2.2 and the relation $P_{t,T}f(x) = \mathbb{E}_{\mathbb{Q}}[f(S_{t,T}^x)]$, after approximation of $x \longmapsto (x - K)^+$ with $\mathcal{C}^2$ functions. $\square$

From the above Proposition 2.3 we recover the formula for the Delta of a European call option in the Black-Scholes model.

Proposition 2.4. *The Delta of a European call option with payoff $F = (S_T - K)^+$ is given by*

$$\eta_t = \Phi \left(\frac{\log(S_t/K) + (\tilde{r}_t + \tilde{\sigma}_t^2/2)(T - t)}{\tilde{\sigma}_t \sqrt{T - t}} \right), \qquad 0 \le t \le T.$$

Proof. By (2.13) we have, taking $x = S_t$,

$$\eta_t = \frac{1}{\sigma_t S_t} e^{-\tilde{r}_t(T-t)} \xi_t$$

$$= e^{-\tilde{r}_t(T-t)} \, \mathbb{E}_{\mathbb{Q}} \left[\frac{S_{t,T}^x}{x} \mathbf{1}_{[K,\infty[}(S_{t,T}^x) \right]$$

$$= e^{-\tilde{r}_t(T-t)}$$

$$\mathbb{E}_{\mathbb{Q}} \left[e^{\tilde{\sigma}_t \hat{B}_{T-t} - \tilde{\sigma}_t^2(T-t)/2 + \tilde{r}_t(T-t)} \mathbf{1}_{[K,\infty[}\left(x e^{\tilde{\sigma}_t \hat{B}_{T-t} - \tilde{\sigma}_t^2(T-t)/2 + \tilde{r}_t(T-t)} \right) \right]$$

$$= \frac{e^{-\tilde{r}_t(T-t)}}{\sqrt{2\pi(T-t)}}$$

$$\int_{\tilde{\sigma}_t(T-t)/2 - \tilde{r}_t(T-t)/\tilde{\sigma}_t + \frac{1}{\tilde{\sigma}_t}\log\frac{K}{x}}^{\infty} e^{\tilde{\sigma}_t y - \tilde{\sigma}_t^2(T-t)/2 + \tilde{r}_t(T-t)} e^{-y^2/(2(T-t))} dy$$

$$= \frac{1}{\sqrt{2\pi(T-t)}} \int_{-d_-/\sqrt{T-t}}^{\infty} e^{-\frac{1}{2(T-t)}(y - \tilde{\sigma}_t(T-t))^2} dy$$

$$= \frac{1}{\sqrt{2\pi(T-t)}} \int_{-d_-/\sqrt{T-t}}^{\infty} e^{-\frac{1}{2(T-t)}(y - \tilde{\sigma}_t(T-t))^2} dy$$

$$= \frac{1}{\sqrt{2\pi}} \int_{-d_+}^{\infty} e^{-\frac{1}{2}y^2} dy$$

$$= \frac{1}{\sqrt{2\pi}} \int_{-\infty}^{d_+} e^{-\frac{1}{2}y^2} dy$$

$$= \Phi(d_+).$$ $\qquad\square$

The result of Proposition 2.4 can also be recovered by (2.7) and direct differentiation of the Black-Scholes function (2.8), as follows:

$$\frac{\partial C}{\partial x}(x,t) = \frac{\partial}{\partial x}\left(x\Phi\left(\frac{\log(x/K) + (\tilde{r}_t + \tilde{\sigma}_t^2/2)(T-t)}{\tilde{\sigma}_t\sqrt{T-t}} \right) \right)$$

$$-K\frac{\partial}{\partial x}\Phi\left(\frac{\log(x/K) + (\tilde{r}_t - \tilde{\sigma}_t^2/2)(T-t)}{\tilde{\sigma}_t\sqrt{T-t}} \right)$$

$$= \Phi\left(\frac{\log(x/K) + (\tilde{r}_t - \tilde{\sigma}_t^2/2)(T-t)}{\tilde{\sigma}_t\sqrt{T-t}} \right)$$

$$+x\frac{\partial}{\partial x}\Phi\left(\frac{\log(x/K) + (\tilde{r}_t + \tilde{\sigma}_t^2/2)(T-t)}{\tilde{\sigma}_t\sqrt{T-t}} \right)$$

$$-K\frac{\partial}{\partial x}\Phi\left(\frac{\log(x/K) + (\tilde{r}_t - \tilde{\sigma}_t^2/2)(T-t)}{\tilde{\sigma}_t\sqrt{T-t}} \right)$$

$$= \Phi\left(\frac{\log(x/K) + (\tilde{r}_t - \tilde{\sigma}_t^2/2)(T-t)}{\tilde{\sigma}_t\sqrt{T-t}} \right)$$

$$+ \frac{1}{\sqrt{2\pi}\tilde{\sigma}_t \sqrt{T-t}} \exp\left(-\frac{1}{2}\left(\frac{\log(x/K) + (\tilde{r}_t + \tilde{\sigma}_t^2/2)(T-t)}{\tilde{\sigma}_t \sqrt{T-t}}\right)^2\right)$$

$$- \frac{K}{\sqrt{2\pi}\tilde{\sigma}_t x \sqrt{T-t}} \exp\left(-\frac{1}{2}\left(\frac{\log(x/K) + (\tilde{r}_t - \tilde{\sigma}_t^2/2)(T-t)}{\tilde{\sigma}_t \sqrt{T-t}}\right)^2\right)$$

$$= \Phi\left(\frac{\log(x/K) + (\tilde{r}_t - \tilde{\sigma}_t^2/2)(T-t)}{\tilde{\sigma}_t \sqrt{T-t}}\right). \tag{2.22}$$

2.6 Exercises

Exercise 2.1. Let $\alpha \in \mathbb{R}$, and consider the solution $\left(X_t^{(\alpha)}\right)_{t\in[0,T]}$ of the stochastic differential equation

$$dX_t^{(\alpha)} = \alpha X_t^{(\alpha)} dt + dB_t, \qquad 0 \le t \le T.$$

(1) Using the Girsanov Theorem 2.1, construct a probability measure $\mathbb{Q}$ under which the process $\left(X_t^{(\alpha)}\right)_{t\in[0,T]}$ becomes a standard Brownian motion.

(2) Compute the expectation

$$\mathbb{E}\left[\exp\left((\beta - \alpha)\int_0^T X_t^{(\alpha)} dX_t^{(\alpha)} + \frac{\alpha^2}{2}\int_0^T \left(X_t^{(\alpha)}\right)^2 dt\right)\right]$$

for all $\beta < 1/T$.

(3) Compute the expectation

$$\mathbb{E}\left[\exp\left(\frac{\alpha^2}{2}\int_0^T (X_t^{(\alpha)})^2 dt\right)\right]$$

for all $\alpha < 1/T$.

Exercise 2.2. Consider the price process $(S_t)_{t\in[0,T]}$ given by

$$\frac{dS_t}{S_t} = \mu dt + \sigma dB_t$$

and a riskless asset of value $A_t = A_0 e^{rt}$, $t \in [0,T]$, with $r > 0$. Let $(\zeta_t, \eta_t)_{t\in[0,T]}$ a self-financing portfolio of value

$$V_t = \eta_t A_t + \zeta_t S_t, \qquad t \in [0,T].$$

(1) Using the Girsanov Theorem 2.1, construct a probability $\mathbb{Q}$ under which the process $\tilde{S}_t := S_t/A_t$, $t \in [0, T]$ is an $\mathcal{F}_t$-martingale.

(2) Compute the arbitrage price

$$C(t, S_t) = e^{-r(T-t)} \, \mathbb{E}_{\mathbb{Q}}[|S_T|^2 | \mathcal{F}_t],$$

at time $t \in [0, T]$, of the contingent claim of payoff $|S_T|^2$.

(3) Compute the portfolio strategy $(\zeta_t, \eta_t)_{t \in [0,T]}$ hedging the claim $|S_T|^2$.

(4) Given $T_0 \in [0, T]$, compute the arbitrage price

$$C(t, S_t) = e^{-r(T-t)} \, \mathbb{E}_{\mathbb{Q}} \left[\frac{S_T}{S_{T_0}} \bigg| \mathcal{F}_t \right],$$

at time $t \in [0, T]$, for the claim of payoff S_T/S_{T_0},

Hint: Consider separately the cases $t \in [0, T_0]$ and $t \in (T_0, T]$.

(5) Compute the portfolio strategy $(\zeta_t, \eta_t)_{t \in [0,T]}$ hedging the claim S_T/S_{T_0}. Check that this strategy is self-financing.

Exercise 2.3.

(1) Solve the stochastic differential equation

$$dS_t = \alpha S_t dt + \sigma dB_t \tag{2.23}$$

in terms of $\alpha, \sigma > 0$, and the initial condition S_0.

(2) For which values α_M of α is the discounted price process $\tilde{S}_t = e^{-rt} S_t$, $t \in [0, T]$, a martingale under $\mathbb{P}$?

(3) Compute the arbitrage price $C(t, S_t) = e^{-r(T-t)} \, \mathbb{E}[\exp(S_T) | \mathcal{F}_t]$ at time $t \in [0, T]$ of the contingent claim of $\exp(S_T)$, with $\alpha = \alpha_M$.

(4) Explicitly compute the strategy $(\zeta_t, \eta_t)_{t \in [0,T]}$ that hedges the contingent claim $\exp(S_T)$.

Chapter 3

Short Term Interest Rate Models

This chapter is a short introduction to some common short term interest rate models. Here we do not aim at completeness as the study of these models has already been extensively developed in the literature, see e.g. [Brigo and Mercurio (2006)], [Carmona and Tehranchi (2006)], [James and Webber (2001)], [Kijima (2003)], [Rebonato (1996)], [Yolcu (2005)]. In the next chapters we will mainly use the Vasicek mean-reverting model in our examples as it allows for explicit calculations.

3.1 Mean-Reverting Models

Interest rates behave differently from stock prices and require the development of specific models to account for properties such as positivity, boundedness, and return to equilibrium.

[Vašiček (1977)] introduced the first model to capture the mean reversion property of interest rates, a property not possessed by geometric Brownian motion. In the Vasicek model, which is based on the Ornstein-Uhlenbeck process, the short term interest rate process $(r_t)_{t \in \mathbb{R}_+}$ solves the equation

$$dr_t = \beta(\alpha - r_t)dt + \sigma dB_t,$$

where $(B_t)_{t \in \mathbb{R}_+}$ is a standard Brownian motion. This model has the interesting properties of being statistically stationary in time in the long run, and to admit a Gaussian $\mathcal{N}(a/b, \sigma^2/(2b))$ invariant distribution when $b > 0$, however its drawback is to allow for negative values of r_t. Explicit formulas for the Vasicek model are obtained in Exercise 1.3 and in Exercise 3.1 below.

The Cox-Ingersoll-Ross (CIR) [Cox *et al.* (1985)] model brings a solution to the positivity problem encountered with the Vasicek model, by the use the nonlinear equation

$$dr_t = \beta(\alpha - r_t)dt + r_t^{1/2}\sigma dB_t.$$

This equation and the properties of its solution are discussed in Exercise 3.2 below.

Other classical mean reverting models include the Courtadon (1982) model

$$dr_t = \beta(\alpha - r_t)dt + \sigma r_t dB_t$$

where α, β, σ are nonnegative, and the exponential-Vasicek model

$$dr_t = r_t(\eta - a \log r_t)dt + \sigma r_t dB_t,$$

where a, η, σ are nonnegative, which is discussed in Exercise 3.1 in this chapter.

More recently, other models preserving the positivity of interest rates have been proposed, cf. eg. [James and Webber (2001)], using stochastic differential equations on manifolds.

3.2   Constant Elasticity of Variance (CEV) Models

Constant Elasticity of Variance models are designed to take into account non-constant volatilities that can vary as a power of the underlying asset. The Marsh-Rosenfeld (1983) model

$$dr_t = (\beta r_t^{-(1-\gamma)} + \alpha r_t)dt + \sigma r_t^{\gamma/2}dB_t$$

where α, β, σ, γ are nonnegative constants, covers most of the CEV models. In particular, for $\beta = 0$ we get the standard CEV model

$$dr_t = \alpha r_t dt + \sigma r_t^{\gamma/2}dB_t,$$

and if $\gamma = 2$ this yields the Dothan [Dothan (1978)], [Pintoux and Privault (2011)] model

$$dr_t = \alpha r_t dt + \sigma r_t dB_t.$$

3.3 Time-Dependent Models

Most of the models discussed in the above sections admit time-dependent extensions. The most elementary example is the Ho-Lee model

$$dr_t = \theta(t)dt + \sigma dB_t,$$

where $\theta(t)$ is a deterministic function of time, which will be used in Exercise 4.1.

The Hull-White model

$$dr_t = (\theta(t) - \alpha(t)r_t)dt + \sigma(t)dB_t$$

is a time-dependent extension of the Vasicek model and will be recovered in Section 6.6. The CIR model also admits a similar time-dependent extension.

Moreover, such time dependent models can be used to fit an initial curve of forward instantaneous rates as in Exercise 8.2-(8), under absence of arbitrage.

The class of short rate interest rate models admits a number of generalizations that can be found in the references quoted in the introduction of this chapter, among which is the class of affine models of the form

$$dr_t = (\eta(t) + \lambda(t)r_t)dt + \sqrt{\delta(t) + \gamma(t)r_t}dB_t. \tag{3.1}$$

Such models are called affine because the associated zero-coupon bonds can be priced using an affine PDE as will be seen in Proposition 4.1 of Chapter 4.

3.4 Exercises

Exercise 3.1. Exponential Vasicek model. Consider a short rate interest rate proces $(r_t)_{t \in \mathbb{R}_+}$ in the exponential Vasicek model:

$$dr_t = r_t(\eta - a \log r_t)dt + \sigma r_t dB_t, \tag{3.2}$$

where η, a, σ are positive parameters.

(1) Find the solution $(Y_t)_{t \in \mathbb{R}_+}$ of the stochastic differential equation
$$dY_t = (\theta - aY_t)dt + \sigma dB_t \qquad (3.3)$$
as a function of the initial condition y_0, where θ, a, σ are positive parameters. *Hint.* Let $Z_t = Y_t - \theta/a$, $t \in \mathbb{R}_+$.

(2) Let $X_t = e^{Y_t}$, $t \in \mathbb{R}_+$. Determine the stochastic differential equation satisfied by $(X_t)_{t \in \mathbb{R}_+}$.

(3) Find the solution $(r_t)_{t \in \mathbb{R}_+}$ of (3.2) in terms of the initial condition r_0.

(4) Compute the conditional mean $\mathbb{E}[r_t | \mathcal{F}_u]$ of r_t, $0 \le u \le t$, where $(\mathcal{F}_u)_{u \in \mathbb{R}_+}$ denotes the filtration generated by the Brownian motion $(B_t)_{t \in \mathbb{R}_+}$.

(5) Compute the conditional variance $\mathrm{Var}[r_t | \mathcal{F}_u] := \mathbb{E}[r_t^2 | \mathcal{F}_u] - (\mathbb{E}[r_t | \mathcal{F}_u])^2$ of r_t, $0 \le u \le t$.

(6) Compute the asymptotic mean and variance $\lim_{t \to \infty} \mathbb{E}[r_t]$ and $\lim_{t \to \infty} \mathrm{Var}[r_t]$.

Exercise 3.2. Cox-Ingerson-Ross model. Consider the equation
$$dr_t = (\alpha - \beta r_t)dt + \sigma\sqrt{r_t}dB_t \qquad (3.4)$$
which models the variations of the short rate process r_t, where α, β, σ and r_0 are positive parameters.

(1) Write down Equation (3.4) in integral form.

(2) Let $u(t) = \mathbb{E}[r_t \mid \mathcal{F}_s]$, $0 \le s \le t$. Show, using the integral form of (3.4), that $u(t)$ satisfies the differential equation
$$u'(t) = \alpha - \beta u(t), \qquad 0 \le s \le t.$$

(3) By an application of Itô's formula to r_t^2, show that
$$dr_t^2 = r_t(2\alpha + \sigma^2 - 2\beta r_t)dt + 2\sigma r_t^{3/2}dB_t. \qquad (3.5)$$

(4) Using the integral form of (3.5), find a differential equation satisfied by $v(t) = E[r_t^2 | \mathcal{F}_s]$, $0 \le s \le t$, and compute $E[r_t^2 | \mathcal{F}_s]$, $0 \le s \le t$. You may assume that $a = 0$ to simplify the computation.
   Hint. The function $f(t) = ce^{-\beta t}/\beta$ solves the differential equation $f'(t) + 2\beta f(t) = ce^{-\beta t}$ for all $c \in \mathbb{R}$.

(5) Let
$$X_t = e^{-\beta t/2}\left(x_0 + \frac{\sigma}{2}\int_0^t e^{\beta s/2}dB_s\right), \qquad t \in \mathbb{R}_+.$$
   Show that X_t satisfies the equation
$$dX_t = \frac{\sigma}{2}dB_t - \frac{\beta}{2}X_t dt.$$

(6) Let $R_t = X_t^2$ and

$$W_t = \int_0^t \text{sign}(X_s)dB_s,$$

where $\text{sign}(x) = 1_{\{x \geq 0\}} - 1_{\{x < 0\}}$, $x \in \mathbb{R}$. Show that

$$dR_t = \left(\frac{\sigma^2}{4} - \beta R_t\right) dt + \sigma\sqrt{R_t}dW_t.$$

Chapter 4

Pricing of Zero-Coupon Bonds

In this chapter we describe the basics of bond pricing in the absence of arbitrage opportunities. Explicit calculations are carried out for the Vasicek model, using both the probabilistic and PDE approaches. The definition of zero-coupon bounds will be used in Chapter 5 in order to construct the forward rate processes.

4.1 Definition and Basic Properties

A zero-coupon bond is a contract priced $P_0(t,T)$ at time $t < T$ to deliver $P_0(T,T) = \$1$ at time T. The computation of the arbitrage price $P_0(t,T)$ of a zero-coupon bond based on an underlying short term interest rate process $(r_t)_{t \in \mathbb{R}_+}$ is a basic and important issue in interest rate modeling.

We may distinguish three different situations:

a) The short rate is a *deterministic* constant $r > 0$.

   In this case, $P_0(t,T)$ should satisfy the equation
   $$e^{r(T-t)} P_0(t,T) = P_0(T,T) = 1,$$
   which leads to
   $$P_0(t,T) = e^{-r(T-t)}, \qquad 0 \leq t \leq T.$$

b) The short rate is a time-dependent and *deterministic* function $(r_t)_{t \in \mathbb{R}_+}$.

   In this case, an argument similar to the above shows that
   $$P_0(t,T) = e^{-\int_t^T r_s ds}, \qquad 0 \leq t \leq T. \tag{4.1}$$

c) The short rate is a *stochastic* process $(r_t)_{t\in\mathbb{R}_+}$.

In this case, formula (4.1) no longer makes sense because the price $P_0(t,T)$, being set at time t, can depend only on information known up to time t. This is in contradiction with (4.1) in which $P_0(t,T)$ depends on the future values of r_s for $s \in [t,T]$.

In the remaining of this chapter we focus on the stochastic case (c). The pricing of the bond $P_0(t,T)$ will follow the following steps, previously used in the case of Black-Scholes pricing.

Pricing bonds with non-zero coupon is not difficult in the case of a deterministic continuous-time coupon yield at rate $c > 0$. In this case the price $P_c(t,T)$ of the coupon bound is given by

$$P_c(t,T) = e^{c(T-t)}P_0(t,T), \qquad 0 \le t \le T.$$

In the sequel we will only consider zero-coupon bonds, and let $P(t,T) = P_0(t,T)$, $0 \le t \le T$.

4.2 Absence of Arbitrage and the Markov Property

Given previous experience with Black-Scholes pricing in Proposition 2.2, it seems natural to write $P(t,T)$ as a conditional expectation under a martingale measure. On the other hand and with respect to point (c) above, the use of conditional expectation appears natural in this framework since it can help us "filter out" the future information past time t contained in (4.1). Thus we postulate that

$$P(t,T) = \mathbb{E}_{\mathbb{Q}}\left[e^{-\int_t^T r_s\,ds}\,\Big|\,\mathcal{F}_t\right] \qquad (4.2)$$

under some martingale (also called risk-neutral) measure $\mathbb{Q}$ yet to be determined. Expression (4.2) makes sense as the "best possible estimate" of the future quantity $e^{-\int_t^T r_s\,ds}$ given information known up to time t.

Assume from now on that the underlying short rate process is solution to the stochastic differential equation

$$dr_t = \mu(t,r_t)dt + \sigma(t,r_t)dB_t \qquad (4.3)$$

where $(B_t)_{t\in\mathbb{R}_+}$ is a standard Brownian motion under $\mathbb{P}$. Recall that for example in the Vasicek model we have

$$\mu(t,x) = a - bx \quad \text{and} \quad \sigma(t,x) = \sigma.$$

Consider a probability measure $\mathbb{Q}$ equivalent to $\mathbb{P}$ and given by its density

$$\frac{d\mathbb{Q}}{d\mathbb{P}} = e^{-\int_0^\infty K_s dB_s - \frac{1}{2}\int_0^\infty |K_s|^2 ds}$$

where $(K_s)_{s\in\mathbb{R}_+}$ is an adapted process satisfying the Novikov integrability condition (2.9). By the Girsanov Theorem 2.1 it is known that

$$\hat{B}_t := B_t + \int_0^t K_s ds$$

is a standard Brownian motion under $\mathbb{Q}$, thus (4.3) can be rewritten as

$$dr_t = \tilde{\mu}(t,r_t)dt + \sigma(t,r_t)d\hat{B}_t$$

where

$$\tilde{\mu}(t,r_t) := \mu(t,r_t) - \sigma(t,r_t)K_t.$$

The process K_t, which is called the "market price of risk", needs to be specified, usually via statistical estimation based on market data.

In the sequel we will assume for simplicity that $K_t = 0$; in other terms we assume that $\mathbb{P}$ is the martingale measure used by the market.

The Markov property states that the future after time t of a Markov process $(X_s)_{s\in\mathbb{R}_+}$ depends only on its present state t and not on the whole history of the process up to time t. It can be stated as follows using conditional expectations:

$$\mathbb{E}[f(X_{t_1},\ldots,X_{t_n}) \mid \mathcal{F}_t] = \mathbb{E}[f(X_{t_1},\ldots,X_{t_n}) \mid X_t]$$

for all times $t_1,\ldots,t_n$ greater than t and all sufficiently integrable function f on $\mathbb{R}^n$, see Appendix A for details.

We will make use of the following fundamental property, cf e.g. Theorem V-32 of [Protter (2005)].

Property 4.1. *All solutions of stochastic differential equations such as (4.3) have the Markov property.*

As a consequence, the arbitrage price $P(t,T)$ satisfies

$$P(t,T) = \mathbb{E}_{\mathbb{Q}}\left[e^{-\int_t^T r_s ds}\Big|\mathcal{F}_t\right]$$

$$= \mathbb{E}_{\mathbb{Q}}\left[e^{-\int_t^T r_s ds}\Big|r_t\right],$$

and depends on r_t only instead of depending on all information available in $\mathcal{F}_t$ up to time t. As such, it becomes a function $F(t,r_t)$ of r_t:

$$P(t,T) = F(t,r_t),$$

meaning that the pricing problem can now be formulated as a search for the function $F(t,x)$.

4.3 Absence of Arbitrage and the Martingale Property

Our goal is now to apply Itô's calculus to $F(t,r_t) = P(t,T)$ in order to derive a PDE satisfied by $F(t,x)$. From Itô's formula Theorem 1.8 we have

$$d\left(e^{-\int_0^t r_s ds}P(t,T)\right) = -r_t e^{-\int_0^t r_s ds}P(t,T)dt + e^{-\int_0^t r_s ds}dP(t,T)$$

$$= -r_t e^{-\int_0^t r_s ds}F(t,r_t)dt + e^{-\int_0^t r_s ds}dF(t,r_t)$$

$$= -r_t e^{-\int_0^t r_s ds}F(t,r_t)dt + e^{-\int_0^t r_s ds}\frac{\partial F}{\partial x}(t,r_t)(\tilde{\mu}(t,r_t)dt + \sigma(t,r_t)d\hat{B}_t)$$

$$+e^{-\int_0^t r_s ds}\left(\frac{1}{2}\sigma^2(t,r_t)\frac{\partial^2 F}{\partial x^2}(t,r_t)dt + \frac{\partial F}{\partial t}(t,r_t)dt\right)$$

$$= e^{-\int_0^t r_s ds}\sigma(t,r_t)\frac{\partial F}{\partial x}(t,r_t)d\hat{B}_t$$

$$+e^{-\int_0^t r_s ds}\left(-r_t F(t,r_t) + \tilde{\mu}(t,r_t)\frac{\partial F}{\partial x}(t,r_t)\right.$$

$$\left.+\frac{1}{2}\sigma^2(t,r_t)\frac{\partial^2 F}{\partial x^2}(t,r_t) + \frac{\partial F}{\partial t}(t,r_t)\right)dt. \qquad (4.4)$$

Next, notice that we have

$$e^{-\int_0^t r_s ds}P(t,T) = e^{-\int_0^t r_s ds}\mathbb{E}_{\mathbb{Q}}\left[e^{-\int_t^T r_s ds}\Big|\mathcal{F}_t\right]$$

$$= \mathbb{E}_{\mathbb{Q}}\left[e^{-\int_0^t r_s ds}e^{-\int_t^T r_s ds}\Big|\mathcal{F}_t\right]$$

$$= \mathbb{E}_{\mathbb{Q}}\left[e^{-\int_0^T r_s ds}\Big|\mathcal{F}_t\right]$$

hence

$$t \longmapsto e^{-\int_0^t r_s ds}P(t,T)$$

is a martingale (see Appendix A) since for any $0 < u < t$ we have:

$$\mathbb{E}_{\mathbb{Q}}\left[e^{-\int_0^t r_s ds} P(t,T)\Big|\mathcal{F}_u\right] = \mathbb{E}_{\mathbb{Q}}\left[\mathbb{E}_{\mathbb{Q}}\left[e^{-\int_0^T r_s ds}\Big|\mathcal{F}_t\right]\Big|\mathcal{F}_u\right]$$

$$= \mathbb{E}_{\mathbb{Q}}\left[e^{-\int_0^T r_s ds}\Big|\mathcal{F}_u\right]$$

$$= \mathbb{E}_{\mathbb{Q}}\left[e^{-\int_0^u r_s ds} e^{-\int_u^T r_s ds}\Big|\mathcal{F}_u\right]$$

$$= e^{-\int_0^u r_s ds}\mathbb{E}_{\mathbb{Q}}\left[e^{-\int_u^T r_s ds}\Big|\mathcal{F}_u\right]$$

$$= e^{-\int_0^u r_s ds} P(u,T).$$

As a consequence, (cf. again Corollary 1, p. 72 of [Protter (2005)]), the above expression (4.4) of

$$d\left(e^{-\int_0^t r_s ds} P(t,T)\right)$$

should contain terms in $d\hat{B}_t$ only, meaning that all terms in dt should vanish inside (4.4). This leads to the identity

$$-r_t F(t,r_t) + \tilde{\mu}(t,r_t)\frac{\partial F}{\partial x}(t,r_t) + \frac{1}{2}\sigma^2(t,r_t)\frac{\partial^2 F}{\partial x^2}(t,r_t) + \frac{\partial F}{\partial t}(t,r_t) = 0,$$

which can be rewritten as in the next proposition.

Proposition 4.1. *The bond pricing PDE for $P(t,T) = F(t,r_t)$ is written as*

$$xF(t,x) = \tilde{\mu}(t,x)\frac{\partial F}{\partial x}(t,x) + \frac{1}{2}\sigma^2(t,x)\frac{\partial^2 F}{\partial x^2}(t,x) + \frac{\partial F}{\partial t}(t,x), \qquad (4.5)$$

subject to the terminal condition

$$F(T,x) = 1. \qquad (4.6)$$

Condition (4.6) is due to the fact that $P(T,T) = \$1$. On the other hand,

$$\left(e^{-\int_0^t r_s ds} P(t,T)\right)_{t\in[0,T]} \qquad \text{and} \qquad (P(t,T))_{t\in[0,T]}$$

respectively satisfy the stochastic differential equations

$$d\left(e^{-\int_0^t r_s ds} P(t,T)\right) = e^{-\int_0^t r_s ds}\sigma(t,r_t)\frac{\partial F}{\partial x}(t,r_t)d\hat{B}_t$$

and

$$dP(t,T) = P(t,T)r_t dt + \sigma(t,r_t)\frac{\partial F}{\partial x}(t,r_t)d\hat{B}_t,$$

i.e.

$$\frac{dP(t,T)}{P(t,T)} = r_t dt + \frac{\sigma(t,r_t)}{P(t,T)}\frac{\partial F}{\partial x}(t,r_t)d\hat{B}_t$$

$$= r_t dt + \sigma(t,r_t)\frac{\partial \log F}{\partial x}(t,r_t)d\hat{B}_t.$$

4.4   PDE Solution: Probabilistic Method

Our goal is now to solve the PDE (4.5) by direct computation of the conditional expectation

$$P(t,T) = \mathbb{E}_{\mathbb{Q}}\left[e^{-\int_t^T r_s ds}\middle|\mathcal{F}_t\right]. \tag{4.7}$$

We will assume that the short rate $(r_t)_{t\in\mathbb{R}_+}$ has the expression

$$r_t = g(t) + \int_0^t h(t,s)dB_s,$$

where $g(t)$ and $h(t,s)$ are deterministic functions, which is the case in particular in the [Vašiček (1977)] model. Letting $u \vee t = \max(u,t)$, using the fact that Wiener integrals are Gaussian random variables (Proposition 1.3), and the Gaussian characteristic function (12.2) and Property (a) of conditional expectations, cf. Appendix A, we have

$$
\begin{aligned}
P(t,T) &= \mathbb{E}_{\mathbb{Q}}\left[e^{-\int_t^T r_s ds}\middle|\mathcal{F}_t\right]\\
&= \mathbb{E}_{\mathbb{Q}}\left[e^{-\int_t^T (g(s)+\int_0^s h(s,u)dB_u)ds}\middle|\mathcal{F}_t\right]\\
&= e^{-\int_t^T g(s)ds}\,\mathbb{E}_{\mathbb{Q}}\left[e^{-\int_t^T \int_0^s h(s,u)dB_u ds}\middle|\mathcal{F}_t\right]\\
&= e^{-\int_t^T g(s)ds}\,\mathbb{E}_{\mathbb{Q}}\left[e^{-\int_0^T \int_{u\vee t}^T h(s,u)dsdB_u}\middle|\mathcal{F}_t\right]\\
&= e^{-\int_t^T g(s)ds}e^{-\int_0^t \int_{u\vee t}^T h(s,u)dsdB_u}\,\mathbb{E}_{\mathbb{Q}}\left[e^{-\int_t^T \int_{u\vee t}^T h(s,u)dsdB_u}\middle|\mathcal{F}_t\right]\\
&= e^{-\int_t^T g(s)ds}e^{-\int_0^t \int_t^T h(s,u)dsdB_u}\,\mathbb{E}_{\mathbb{Q}}\left[e^{-\int_t^T \int_u^T h(s,u)dsdB_u}\middle|\mathcal{F}_t\right]\\
&= e^{-\int_t^T g(s)ds}e^{-\int_0^t \int_t^T h(s,u)dsdB_u}\,\mathbb{E}_{\mathbb{Q}}\left[e^{-\int_t^T \int_u^T h(s,u)dsdB_u}\right]\\
&= e^{-\int_t^T g(s)ds}e^{-\int_0^t \int_t^T h(s,u)dsdB_u}e^{\frac{1}{2}\int_t^T \left(\int_u^T h(s,u)ds\right)^2 du}.
\end{aligned}
$$

Recall that in the [Vašiček (1977)] model, i.e. when the short rate process is solution of

$$dr_t = (a - br_t)dt + \sigma dB_t,$$

and the market price of risk is $K_t = 0$, we have the explicit solution, cf. Exercise 1.3 and Exercise 3.1:

$$r_t = r_0 e^{-bt} + \frac{a}{b}(1 - e^{-bt}) + \sigma \int_0^t e^{-b(t-s)}dB_s, \tag{4.8}$$

hence the above calculation yields

$$P(t,T) = \mathbb{E}_{\mathbb{Q}}\left[e^{-\int_t^T r_s ds}\,\Big|\,\mathcal{F}_t\right]$$

$$= e^{-\int_t^T (r_0 e^{-bs} + \frac{a}{b}(1-e^{-bs}))ds}\, e^{-\sigma \int_0^t \int_t^T e^{-b(s-u)}ds\,dB_u}$$

$$\times e^{\frac{\sigma^2}{2}\int_t^T \left(\int_u^T e^{-b(s-u)}ds\right)^2 du}$$

$$= e^{-\int_t^T (r_0 e^{-bs} + \frac{a}{b}(1-e^{-bs}))ds}\, e^{-\frac{\sigma}{b}(1-e^{-b(T-t)})\int_0^t e^{-b(t-u)}dB_u}$$

$$\times e^{\frac{\sigma^2}{2}\int_t^T e^{2bu}\left(\frac{e^{-bu}-e^{-bT}}{b}\right)^2 du}$$

$$= e^{-\frac{r_t}{b}(1-e^{-b(T-t)}) + \frac{1}{b}(1-e^{-b(T-t)})(r_0 e^{-bt} + \frac{a}{b}(1-e^{-bt}))}$$

$$\times e^{-\int_t^T (r_0 e^{-bs} + \frac{a}{b}(1-e^{-bs}))ds + \frac{\sigma^2}{2}\int_t^T e^{2bu}\left(\frac{e^{-bu}-e^{-bT}}{b}\right)^2 du}$$

$$= e^{C(T-t)r_t + A(T-t)},$$

where

$$C(T-t) = -\frac{1}{b}(1 - e^{-b(T-t)}),$$

and

$$A(T-t) = \frac{1}{b}(1 - e^{-b(T-t)})(r_0 e^{-bt} + \frac{a}{b}(1-e^{-bt}))$$

$$- \int_t^T (r_0 e^{-bs} + \frac{a}{b}(1-e^{-bs}))ds$$

$$+ \frac{\sigma^2}{2}\int_t^T e^{2bu}\left(\frac{e^{-bu}-e^{-bT}}{b}\right)^2 du$$

$$= \frac{1}{b}(1 - e^{-b(T-t)})(r_0 e^{-bt} + \frac{a}{b}(1-e^{-bt}))$$

$$- \frac{r_0}{b}(e^{-bt} - e^{-bT}) - \frac{a}{b}(T-t) + \frac{a}{b^2}(e^{-bt} - e^{-bT})$$

$$+ \frac{\sigma^2}{2b^2}\int_t^T \left(1 + e^{-2b(T-u)} - 2e^{-b(T-u)}\right)du$$

$$= \frac{a}{b^2}(1 - e^{-b(T-t)})(1 - e^{-bt}) - \frac{a}{b}(T-t) + \frac{a}{b^2}(e^{-bt} - e^{-bT})$$

$$+ \frac{\sigma^2}{2b^2}(T-t) + \frac{\sigma^2}{2b^2}e^{-2bT}\int_t^T e^{2bu}du - \frac{\sigma^2}{b^2}e^{-bT}\int_t^T e^{bu}du$$

$$= \frac{a}{b^2}(1 - e^{-b(T-t)}) + \frac{\sigma^2 - 2ab}{2b^2}(T-t)$$

$$+ \frac{\sigma^2}{4b^3}(1 - e^{-2b(T-t)}) - \frac{\sigma^2}{b^3}(1 - e^{-b(T-t)})$$

$$= \frac{4ab - 3\sigma^2}{4b^3} + \frac{\sigma^2 - 2ab}{2b^2}(T-t)$$

$$+ \frac{\sigma^2 - ab}{b^3}e^{-b(T-t)} - \frac{\sigma^2}{4b^3}e^{-2b(T-t)}.$$

See Exercise 4.5 for another way to calculate $P(t, T)$ in the [Vašiček (1977)] model.

Note that more generally, all affine short rate models as defined in Relation (3.1), including the Vasicek model, will yield a bond pricing formula of the form

$$P(t, T) = e^{A(T-t)+C(T-t)r_t},$$

cf. e.g. § 3.2.4. of [Brigo and Mercurio (2006)].

4.5   PDE Solution: Analytical Method

In this section we still assume that the underlying short rate process is the Vasicek process solution of (4.3). In order to solve the PDE (4.5) analytically we look for a solution of the form

$$F(t, x) = e^{A(T-t)+xC(T-t)}, \qquad (4.9)$$

where A and C are functions to be determined under the conditions $A(0) = 0$ and $C(0) = 0$. Plugging (4.9) into the PDE (4.5) yields the system of Riccati and linear differential equations

$$\begin{cases} -A'(s) = -aC(s) - \dfrac{\sigma^2}{2}C^2(s) \\[2mm] -C'(s) = bC(s) + 1, \end{cases}$$

which can be solved to recover

$$A(s) = \frac{4ab - 3\sigma^2}{4b^3} + s\frac{\sigma^2 - 2ab}{2b^2} + \frac{\sigma^2 - ab}{b^3}e^{-bs} - \frac{\sigma^2}{4b^3}e^{-2bs}$$

and

$$C(s) = -\frac{1}{b}(1 - e^{-bs}).$$

As a verification we easily check that $C(s)$ and $A(s)$ given above do satisfy

$$bC(s) + 1 = -e^{-bs} = -C'(s),$$

and

$$\begin{aligned} aC(s) + \frac{\sigma^2 C^2(s)}{2} &= -\frac{a}{b}(1 - e^{-bs}) + \frac{\sigma^2}{2b^2}(1 - e^{-bs})^2 \\[2mm] &= \frac{\sigma^2 - 2ab}{2b^2} - \frac{\sigma^2 - ab}{b^2}e^{-bs} + \frac{\sigma^2}{2b^2}e^{-2bs} \\[2mm] &= A'(s). \end{aligned}$$

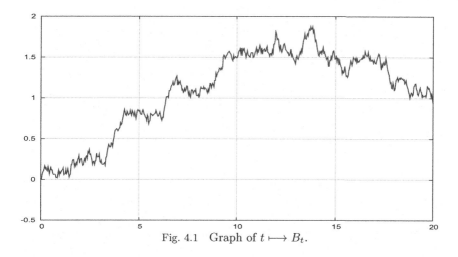

Fig. 4.1   Graph of $t \longmapsto B_t$.

4.6   Numerical Simulations

Given the Brownian path represented in Figure 4.1, Figure 4.2 presents the corresponding random simulation of $t \longmapsto r_t$ in the Vasicek model with $r_0 = a/b = 5\%$, i.e. the reverting property of the process is with respect to its initial value $r_0 = 5\%$. Note that the interest rate in Figure 4.2 becomes negative for a short period of time, which is unusual for interest rates but may nevertheless happen [Bass (October 7, 2007)].

Fig. 4.2   Graph of $t \longmapsto r_t$.

Figure 4.3 presents a random simulation of $t \longmapsto P(t,T)$ in the same Vasicek model. The graph of the corresponding deterministic bond price obtained for $a = b = \sigma = 0$ is also shown on the same Figure 4.3.

Fig. 4.3   Graphs of $t \longmapsto P(t,T)$ and $t \longmapsto e^{-r_0(T-t)}$.

Figure 4.4 presents a random simulation of $t \longmapsto P(t,T)$ for a coupon bond with price $P_c(t,T) = e^{c(T-t)}P(t,T)$, $0 \le t \le T$.

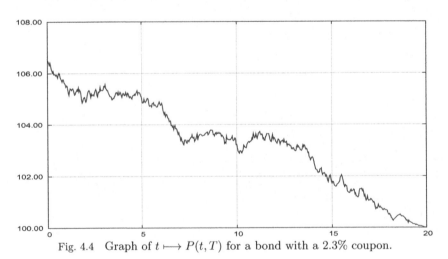

Fig. 4.4   Graph of $t \longmapsto P(t,T)$ for a bond with a 2.3% coupon.

Finally we consider the graphs of the functions A and C in Figures 4.5 and 4.6 respectively.

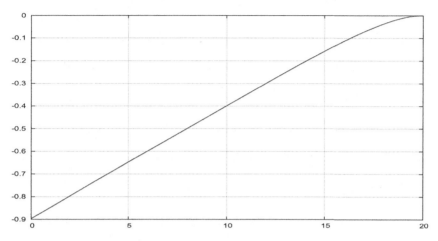

Fig. 4.5 Graph of $t \longmapsto A(T - t)$.

Fig. 4.6 Graph of $t \longmapsto C(T - t)$.

The solution of the pricing PDE, which can be useful for calibration purposes, is represented in Figure 4.7.

Fig. 4.7   Graph of $(x, t) \longmapsto \exp(A(T - t) + xC(T - t))$.

4.7   Exercises

Exercise 4.1. Consider a short term interest rate process $(r_t)_{t \in \mathbb{R}_+}$ in a Ho-Lee model with constant coefficients:

$$dr_t = \theta dt + \sigma dW_t,$$

and let $P(t, T)$ will denote the arbitrage price of a zero-coupon bond in this model:

$$P(t, T) = \mathbb{E}_\mathbb{P} \left[\exp \left(-\int_t^T r_s ds \right) \bigg| \mathcal{F}_t \right], \qquad 0 \le t \le T. \qquad (4.10)$$

(1) State the bond pricing PDE satisfied by the function $F(t, x)$ defined via

$$F(t, x) = \mathbb{E}_\mathbb{P} \left[\exp \left(-\int_t^T r_s ds \right) \bigg| r_t = x \right], \qquad 0 \le t \le T.$$

(2) Compute the arbitrage price $F(t, r_t) = P(t, T)$ from its expression (4.10) as a conditional expectation.
(3) Check that the function $F(t, x)$ computed in Question (2) does satisfy the PDE derived in Question (1).

Exercise 4.2. (Exercise 3.2 continued). Write down the bond pricing PDE for the function

$$F(t, x) = E \left[e^{-\int_t^T r_s ds} \bigg| r_t = x \right]$$

and show that in case $\alpha = 0$ the corresponding bond price $P(t,T)$ equals

$$P(t,T) = e^{-B(T-t)r_t}, \qquad 0 \leq t \leq T,$$

where

$$B(x) = \frac{2(e^{\gamma x} - 1)}{2\gamma + (\beta + \gamma)(e^{\gamma x} - 1)},$$

with $\gamma = \sqrt{\beta^2 + 2\sigma^2}$.

Exercise 4.3. Let $(r_t)_{t \in \mathbb{R}_+}$ denote a short term interest rate process. For any $T > 0$, let $P(t,T)$ denote the price at time $t \in [0,T]$ of a zero coupon bond defined by the stochastic differential equation

$$\frac{dP(t,T)}{P(t,T)} = r_t dt + \sigma_t^T dB_t, \qquad 0 \leq t \leq T, \qquad (4.11)$$

under the terminal condition $P(T,T) = 1$, where $(\sigma_t^T)_{t \in [0,T]}$ is an adapted process. Let the forward measure $\mathbb{P}_T$ be defined by

$$\mathbb{E}\left[\frac{d\mathbb{P}_T}{d\mathbb{P}}\,\middle|\,\mathcal{F}_t\right] = \frac{P(t,T)}{P(0,T)} e^{-\int_0^t r_s ds}, \qquad 0 \leq t \leq T.$$

Recall that

$$B_t^T := B_t - \int_0^t \sigma_s^T ds, \qquad 0 \leq t \leq T,$$

is a standard Brownian motion under $\mathbb{P}_T$.

(1) Solve the stochastic differential equation (4.11).
(2) Derive the stochastic differential equation satisfied by the discounted bond price process

$$t \longmapsto e^{-\int_0^t r_s ds} P(t,T), \qquad 0 \leq t \leq T,$$

and show that it is a martingale.
(3) Show that

$$\mathbb{E}\left[e^{-\int_0^T r_s ds}\,\middle|\,\mathcal{F}_t\right] = e^{-\int_0^t r_s ds} P(t,T), \qquad 0 \leq t \leq T.$$

(4) Show that

$$P(t,T) = \mathbb{E}\left[e^{-\int_t^T r_s ds}\,\middle|\,\mathcal{F}_t\right], \qquad 0 \leq t \leq T.$$

(5) Compute $P(t,S)/P(t,T)$, $0 \leq t \leq T$, show that it is a martingale under $\mathbb{P}_T$ and that

$$P(T,S) = \frac{P(t,S)}{P(t,T)} \exp\left(\int_t^T (\sigma_s^S - \sigma_s^T) dB_s^T - \frac{1}{2}\int_t^T (\sigma_s^S - \sigma_s^T)^2 ds\right).$$

Exercise 4.4. (Exercise 1.8 continued). Assume that the price $P(t, T)$ of a zero coupon bond is modeled as

$$P(t, T) = e^{-\mu(T-t)+X_t^T}, \qquad t \in [0, T],$$

where $\mu > 0$. Show that the terminal condition $P(T, T) = 1$ is satisfied.

Problem 4.5. Consider the stochastic differential equation

$$\begin{cases} dX_t = -bX_t dt + \sigma dB_t, & t > 0, \\ \\ X_0 = 0, \end{cases} \qquad (4.12)$$

where b and σ are positive parameters and $(B_t)_{t \in \mathbb{R}_+}$ is a standard Brownian motion under $\mathbb{P}$, generating the filtration $(\mathcal{F}_t)_{t \in \mathbb{R}_+}$. Let the short term interest rate process $(r_t)_{t \in \mathbb{R}_+}$ be given by

$$r_t = r + X_t, \qquad t \in \mathbb{R}_+,$$

where $r > 0$ is a given constant. Recall that from the Markov property, the arbitrage price

$$P(t, T) = \mathbb{E}_{\mathbb{P}} \left[\exp \left(-\int_t^T r_s ds \right) \Big| \mathcal{F}_t \right], \qquad 0 \le t \le T,$$

of a zero-coupon bond is a function $F(t, X_t) = P(t, T)$ of t and X_t.

(1) Using Itô's calculus, derive the PDE satisfied by the function $(t, x) \longmapsto F(t, x)$.

(2) Solve the stochastic differential equation (4.12).

(3) Show that

$$\int_0^t X_s ds = -\frac{\sigma}{b} \left(\int_0^t (e^{-b(t-s)} - 1) dB_s \right), \qquad t > 0.$$

(4) Show that for all $0 \le t \le T$,

$$\int_t^T X_s ds = -\frac{\sigma}{b} \left(\int_0^t (e^{-b(T-s)} - e^{-b(t-s)}) dB_s + \int_t^T (e^{-b(T-s)} - 1) dB_s \right).$$

(5) Show that

$$\mathbb{E} \left[\int_t^T X_s ds \Big| \mathcal{F}_t \right] = -\frac{\sigma}{b} \int_0^t (e^{-b(T-s)} - e^{-b(t-s)}) dB_s.$$

(6) Show that

$$\mathbb{E}\left[\int_t^T X_s ds \Big| \mathcal{F}_t\right] = \frac{X_t}{b}(1 - e^{-b(T-t)}).$$

(7) Show that

$$\mathrm{Var}\left[\int_t^T X_s ds \Big| \mathcal{F}_t\right] = \frac{\sigma^2}{b^2}\int_t^T (e^{-b(T-s)} - 1)^2 ds.$$

(8) What is the distribution of $\int_t^T X_s ds$ given $\mathcal{F}_t$?

(9) Compute the arbitrage price $P(t,T)$ from its expression (4.10) as a conditional expectation and show that

$$P(t,T) = e^{A(t,T)-r(T-t)+X_t C(t,T)},$$

where $C(t,T) = (e^{-b(T-t)} - 1)/b$ and

$$A(t,T) = \frac{\sigma^2}{2b^2}\int_t^T (e^{-b(T-s)} - 1)^2 ds.$$

(10) Check explicitly that the function $F(t,x) = e^{A(t,T)+r(T-t)+xC(t,T)}$ computed in Question (9) does solve the PDE derived in Question (1).

Chapter 5

Forward Rate Modeling

In this chapter we define the forward and instantaneous forward rates from absence of arbitrage arguments, and using the construction of zero-coupon presented in Chapter 4. We also consider the problem of parametrization of forward rates.

5.1 Forward Contracts

Financial institutions often require the possibility to agree at a present time t for a loan to be delivered over a future period of time $[T, S]$ at a rate $r(t, T, S)$, $t \leq T \leq S$. This type of forward interest rate contracts gives its holder a loan decided at present time t over a future period of time $[T, S]$. In other words, at time t an investor applies for a loan on the period $[T, S]$, in order to repay a unit amount at time S.

The interest rate to be applied to this loan is denoted by $f(t, T, S)$ and is called a forward rate. Here we are interested in determining the arbitrage or "fair" value of this rate using the instruments available in a bond market, that is bonds priced at $P(t, T)$ for various maturity dates $T > t$.

The loan can be realized using the bonds available on the market by proceeding in two steps:

1) at time t, borrow \$1 at the price $P(t, S)$, to be repaid at time S.

2) since one only needs the money at time T, it makes sense to invest the amount $P(t, S)$ over the period $[t, T]$ in a bond with maturity T, that will yield $P(t, S)/P(t, T)$ at time T.

As a consequence the investor will receive $P(t,S)/P(t,T)$ at time T and repay a unit amount \$1 at time S.

Using exponential compounding, the corresponding forward rate $f(t,T,S)$, $0 \leq t \leq T \leq S$, is then given by the relation

$$\exp\left((S-T)f(t,T,S)\right) = \left(\frac{P(t,S)}{P(t,T)}\right)^{-1},$$

which leads to the following definition. Here we used exponential compounding of interest rates, however linear compounding could also be used, as will be seen when dealing with LIBOR rates in Section 10.4 of Chapter 10.

Definition 5.1. *The forward rate $f(t,T,S)$ at time t for a loan on $[T,S]$ is given by*

$$f(t,T,S) = -\frac{\log P(t,S) - \log P(t,T)}{S-T}.$$

The *spot* forward rate $F(t,T)$ is given by

$$F(t,T) := f(t,t,T) = -\frac{\log P(t,T)}{T-t}.$$

Figure 5.1 presents a typical forward rate curve on the LIBOR (London Interbank Offered Rate) market with $t = 07$ may 2003, $\delta =$ six months.

TimeSerieNb	505
AsOfDate	7-mai-03
2D	2,55
1W	2,53
1M	2,56
2M	2,52
3M	2,48
1Y	2,34
2Y	2,49
3Y	2,79
4Y	3,07
5Y	3,31
6Y	3,52
7Y	3,71
8Y	3,88
9Y	4,02
10Y	4,14
11Y	4,23
12Y	4,33
13Y	4,4
14Y	4,47
15Y	4,54
20Y	4,74
25Y	4,83
30Y	4,86

Fig. 5.1   Graph of $T \longmapsto f(t,T,T+\delta)$.

Recall that in the Vasicek model, i.e. when the short rate process is solution of

$$dr_t = (a - br_t)dt + \sigma dB_t$$

we have

$$P(t, T) = e^{C(T-t)r_t + A(T-t)}$$

where

$$C(T - t) = -\frac{1}{b}(1 - e^{-b(T-t)})$$

and

$$A(T - t) = \frac{4ab - 3\sigma^2}{4b^3} + \frac{\sigma^2 - 2ab}{2b^2}(T - t) + \frac{\sigma^2 - ab}{b^3}e^{-b(T-t)} - \frac{\sigma^2}{4b^3}e^{-2b(T-t)},$$

cf. Chapter 4, hence

$$\log P(t, T) = A(T - t) + r_t C(T - t)$$

and

$$
\begin{aligned}
f(t, T, S) &= -\frac{\log P(t, S) - \log P(t, T)}{S - T} \\
&= -\frac{r_t(C(S - t) - C(T - t)) + A(S - t) - A(T - t))}{S - T} \\
&= -\frac{\sigma^2 - 2ab}{2b^2} \\
&\quad - \frac{1}{S - T}\left(\left(\frac{r_t}{b} + \frac{\sigma^2 - ab}{b^3}\right)(e^{-b(S-t)} - e^{-b(T-t)}) \right. \\
&\quad \left. - \frac{\sigma^2}{4b^3}(e^{-2b(S-t)} - e^{-2b(T-t)})\right).
\end{aligned}
$$

In this model the forward rate $t \longmapsto f(t, T, S)$ can be represented as in Figure 5.2, with here $b/a > r_0$.

Note that the forward rate cure $t \longmapsto f(t, T, S)$ is flat for small values of t. This modelization issue will be reconsidered in the framework of multifactor models in Chapter 8.

Fig. 5.2   Forward rate process $t \longmapsto f(t, T, S)$.

5.2   Instantaneous Forward Rate

The instantaneous forward rate $f(t, T)$ is defined by taking the limit of $f(t, T, S)$ as $S \searrow T$, i.e.

$$
\begin{aligned}
f(t, T) : & = -\lim_{S \searrow T} \frac{\log P(t, S) - \log P(t, T)}{S - T} \\
& = -\lim_{\varepsilon \searrow 0} \frac{\log P(t, T + \varepsilon) - \log P(t, T)}{\varepsilon} \\
& = -\frac{\partial \log P(t, T)}{\partial T} \\
& = -\frac{1}{P(t, T)} \frac{\partial P(t, T)}{\partial T}.
\end{aligned}
$$

The above equation can be viewed as a differential equation to be solved for $\log P(t, T)$ under the initial condition $P(T, T) = 1$, which gives

$$
\begin{aligned}
\log P(t, T) & = \log P(t, T) - \log P(t, t) \\
& = \int_t^T \frac{\partial \log P(t, s)}{\partial s} ds \\
& = -\int_t^T f(t, s) ds,
\end{aligned}
$$

hence

$$P(t,T) = \exp\left(-\int_t^T f(t,s)ds\right), \qquad 0 \le t \le T. \qquad (5.1)$$

The forward rate $f(t,T,S)$ can be recovered from the instantaneous forward rate $f(t,s)$, as:

$$f(t,T,S) = \frac{1}{S-T}\int_T^S f(t,s)ds, \qquad 0 \le t \le T < S. \qquad (5.2)$$

Note that when the short rate $(r_s)_{s\in\mathbb{R}_+}$ is a deterministic function we have

$$P(t,T) = \exp\left(-\int_t^T f(t,s)ds\right) = \exp\left(-\int_t^T r_s ds\right), \qquad (5.3)$$

$0 \le t \le T$, hence the instantaneous forward rate $f(t,T)$ is also deterministic and independent of t:

$$f(t,T) = r_T, \qquad 0 \le t \le T,$$

and the forward rate $f(t,T,S)$ is given by

$$f(t,T,S) = \frac{1}{S-T}\int_T^S r_s ds, \qquad 0 \le t \le T < S,$$

which is the average of the deterministic interest rate r_s over the time period $[T,S]$.

Furthermore, in case $(r_s)_{s\in\mathbb{R}_+}$ is time-independent and equal to a constant value $r > 0$, all rates coincide and become equal to r:

$$r_s = f(t,s) = f(t,T,S) = r, \qquad 0 \le t \le T \le s < S.$$

For example, in the Vasicek stochastic interest rate model considered in Section 5.1 we have

$$f(t,T) := -\frac{\partial \log P(t,T)}{\partial T} \qquad (5.4)$$

$$= r_t e^{-b(T-t)} + \frac{a}{b}(1 - e^{-b(T-t)}) - \frac{\sigma^2}{2b^2}(1 - e^{-b(T-t)})^2,$$

$0 \le t \le T$.

In this model the instantaneous forward rate $t \longmapsto f(t,T)$ can be represented as in Figure 5.3, with here $t = 0$ and $b/a > r_0$. On the other hand, the instantaneous forward rate $T \longmapsto f(t,T)$ can be represented as in Figure 5.4, with here $t = 0$ and $b/a > r_0$.

Fig. 5.3   Instantaneous forward rate process $t \longmapsto f(t,T)$.

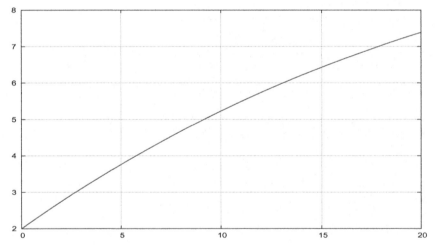

Fig. 5.4   Instantaneous forward rate process $T \longmapsto f(0,T)$.

5.3   Short Rates

The underlying short term interest rate process $(r_t)_{t \in \mathbb{R}_+}$ can be recovered from its relation to the bond price

$$P(t,T) = \mathbb{E}\left[\exp\left(-\int_t^T r_s ds\right)\Big| r_t\right].$$

Indeed we have

$$\frac{\partial P}{\partial T}(t, T) = \frac{\partial}{\partial T} \mathbb{E}\left[\exp\left(-\int_t^T r_s ds\right)\bigg| r_t\right]$$

$$= \mathbb{E}\left[\frac{\partial}{\partial T}\exp\left(-\int_t^T r_s ds\right)\bigg| r_t\right]$$

$$= -\mathbb{E}\left[r_T\exp\left(-\int_t^T r_s ds\right)\bigg| r_t\right],$$

hence

$$\lim_{T\searrow t}\frac{\partial P}{\partial T}(t, T) = -\mathbb{E}[r_t|r_t] = -r_t,$$

and the limit $\lim_{T\searrow t} f(t, T)$ of instantaneous forward rates equals the short rate r_t, i.e. we have

$$\lim_{T\searrow t} f(t, T) = -\lim_{T\searrow t}\frac{1}{P(t, T)}\frac{\partial P(t, T)}{\partial T} = r_t,$$

since $\lim_{T\searrow t} P(t, T) = 1$.

Note that in the Vasicek model the relation $\lim_{T\searrow t} f(t, T) = r_t$ is easy to recover from Relation (5.4).

5.4 Parametrization of Forward Rates

The forward rate curve represented in Figure 5.4 has some similarities with the market data curve of Figure 5.1 (for instance, it is increasing), however it clearly misses some features traditionally observed in forward curves, such as the hump at the left hand side of the graph. For this reason, other parametrizations of forward rates have been introduced.

In the sequel we will frequently use the Musiela convention, i.e. we will write

$$g(x) = f(t, t + x) = f(t, T),$$

under the substitution $x = T - t$, $x \geq 0$.

Nelson-Siegel parametrization

This family of curves is parametrized by 4 coefficients z_1, z_2, z_3, z_4, as

$$g(x) = z_1 + (z_2 + z_3 x)e^{-x z_4}, \qquad x \geq 0.$$

An example of a graph obtained by the Nelson-Siegel parametrization is given in Figure 5.5, for $z_1 = 1$, $z_2 = -10$, $z_3 = 100$, $z_4 = 10$.

Fig. 5.5 Graph of $x \longmapsto g(x)$ in the Nelson-Siegel model.

Svensson parametrization

The advantage of this family of curves is the ability to reproduce two humps instead of one, the location and height of which can be chosen via 6 parameters z_1, z_2, z_3, z_4, z_5, z_6 as

$$g(x) = z_1 + (z_2 + z_3 x)e^{-x z_4} + z_5 x e^{-x z_6}, \qquad x \geq 0.$$

A typical graph of a Svensson parametrization is given in Figure 5.6, for $z_1 = 7$, $z_2 = -5$, $z_3 = -100$, $z_4 = 10$, $z_5 = -1/2$, $z_6 = -1$.

5.5 Curve Estimation

A simple way to estimate the forward curve based on market data of bond prices $(P(t, T_k))_{k=1,\dots,n}$ with maturities $T_1, T_2, \dots, T_n$ is to assume that the instantaneous forward rate is a step function:

$$f(t, x) = g(x) = \sum_{k=1}^{n} \alpha_k 1_{(T_{k-1}, T_k]}(x),$$

with $T_0 = 0$. In this case we have the relation

$$\frac{P(t, T_k)}{P(t, T_{k-1})} = \exp\left(-\int_{T_{k-1}}^{T_k} g(x)dx\right) = \exp\left(-\alpha_k(T_k - T_{k-1})\right),$$

hence

$$\alpha_k = -\frac{1}{T_k - T_{k-1}} \log \frac{P(t, T_k)}{P(t, T_{k-1})}, \qquad k = 1, \dots, n.$$

Fig. 5.6 Graph of $x \longmapsto g(x)$ in the Svensson model.

A more realistic estimation can be made by requiring some additional smoothness on $g(x)$, for example requiring that it be twice differentiable. In this case the estimate $g(x)$ will be the outcome of the minimization problem

$$
\min_{g} \left(\lambda \int_0^{T_n} |g''(x)|^2 dx + \sum_{k=1}^{n} \beta_k \left| \frac{P(t, T_k)}{P(t, T_{k-1})} - \exp\left(-\int_{T_{k-1}}^{T_k} g(x) dx \right) \right|^2 \right)
$$

where $\beta_1, \ldots, \beta_n$ and λ are positive coefficients.

5.6 Exercises

Exercise 5.1. (Exercise 4.1 continued).

(1) Compute the forward rate $f(t, T, S)$ in this model.
(2) Compute the instantaneous forward rate $f(t, T)$ in this model.

Exercise 5.2. (Exercise 4.5 continued).

(1) Compute the forward rate

$$
f(t, T, S) = -\frac{\log P(t, S) - \log P(t, T)}{S - T}.
$$

(2) Compute the instantaneous forward rate

$$f(t,T) = \lim_{S \searrow T} f(t,T,S).$$

Exercise 5.3. (Exercise 4.3 continued).

(1) Compute the forward rate

$$f(t,T,S) = -\frac{1}{S-T}(\log P(t,S) - \log P(t,T)).$$

(2) Compute the instantaneous forward rate

$$f(t,T) = -\lim_{S \searrow T} \frac{1}{S-T}(\log P(t,S) - \log P(t,T)).$$

(3) Show that the limit $\lim_{T \searrow t} f(t,T)$ does not exist in $L^2(\Omega)$.

(4) Show that $P(t,T)$ satisfies the stochastic differential equation

$$\frac{dP(t,T)}{P(t,T)} = \sigma dB_t + \frac{1}{2}\sigma^2 dt - \frac{\log P(t,T)}{T-t}dt, \qquad t \in [0,T].$$

(5) Show, using the results of Exercise 4.3-(4), that

$$P(t,T) = \mathbb{E}\left[e^{-\int_t^T r_s^T ds}\middle| \mathcal{F}_t\right],$$

where $(r_t^T)_{t \in [0,T]}$ is a process to be determined.

Chapter 6

The Heath-Jarrow-Morton (HJM) Model

In this chapter we present the general framework of Heath, Jarrow and Morton [Heath *et al.* (1992)] in which the evolution of forward rates is infinite dimensional and viewed as a stochastic process valued in a function space. We also expose the HJM absence of arbitrage condition. Under this condition we recover some time dependent short term interest rate models such as the Hull-White model.

6.1 Restatement of Objectives

Before proceeding further we would like to recall our general objectives, which are:

(1) to find a (stochastic) model for the underlying (forward) interest rate processes.

(2) to derive (option) pricing formulas as functions of the model's parameters.

(3) to calibrate the parameters of the model by matching computed prices to market data.

(4) to compute "new" prices using the calibrated pricing formulas.

What type of option can we consider on interest rates?

Caps and floors are standard examples of options on interest rates. An interest rate cap will protect a borrower against interest rates going above a certain level κ. As an example, a cap on the underlying short rate at time T yields a payoff equal to

$$r_T - \min(\kappa, r_T) = (r_T - \kappa)^+$$

expressed in interest rate (base) points.

However this type of cap makes little sense in practice, since:

a) the rate r_T is not a tradable asset,

b) r_T is an instantaneous rate which makes sense only over an infinitesimally short period of time $[T, T + dt]$.

With reference to point (a) above, the bond price $P(T, S)$ is a tradable asset and in a Markovian setting it can be written as a function

$$P(T, S) = F(T, r_T)$$

of r_T, thus a put option with payoff

$$(K - P(T, S))^+ = (K - F(T, r_T))^+$$

does certainly make sense.

Concerning point (b), one could consider a floor on the average of the short rate over a given period of time, with payoff

$$\max \left(\kappa, \frac{1}{S - T} \int_T^S r_s ds \right),$$

however this average requires the knowledge of information up to time S and it is not directly connected to the bond price, except in case the short rate is deterministic.

Recall that in case $(r_t)_{t \in \mathbb{R}_+}$ is *deterministic*, this average is equal to the forward rate

$$f(t, T, S) = \frac{1}{S - T} \int_T^S r_s ds,$$

cf. Relation (5.3), and that in the general case the forward rate can be obtained by averaging the instantaneous rate $f(t, s)$ as

$$f(t, T, S) = \frac{1}{S - T} \int_T^S f(t, s) ds,$$

cf. Relation (5.2).

In practice, interest rate option contracts are made on forward rates rather than on the short rate process $(r_t)_{t \in \mathbb{R}_+}$. In addition, since $f(t, T, S)$ is part of the data known at time t (i.e. it is $\mathcal{F}_t$-measurable), it makes more sense to consider a cap on the *forward spot rate*

$$f(T, T, S) = \frac{1}{S - T} \int_T^S f(T, s) ds$$

which appears as being random at time t (precisely, it is $\mathcal{F}_T$-measurable but not $\mathcal{F}_t$-measurable).

Thus an interest rate floor will generally yield a payoff of the form.

$$\max(\kappa, f(T, T, S)) = \kappa + (f(T, T, S) - \kappa)^+.$$

Under a different choice of payoff function, the payoff of the contract can take the form

$$\left(K - e^{-(S-T)f(T,T,S)} \right)^+ = \left(K - \exp\left(-\int_T^S f(T, s) ds \right) \right)^+$$

$$= (K - P(T, S))^+,$$

and we recover the standard European put option on the bond price $P(T, S) = e^{-(S-T)f(T,T,S)}$.

This leads us to the important question of how to model the forward rate $f(t, T, S)$; precisely in the next section we will start by considering the modeling of the instantaneous forward rate $f(t, T)$.

The graph given in Figure 6.1 presents a possible random evolution of a forward interest rate curve using the Musiela convention, i.e. for all $t \in \mathbb{R}_+$ a sample of the instantaneous forward curve $x \longmapsto f(t, t+x)$ is represented.

6.2 Forward Vasicek Rates

Recall that in the Vasicek model, the instantaneous forward rate process (5.4) is given by

$$f(t, T) = r_t e^{-b(T-t)} + \frac{a}{b}(1 - e^{-b(T-t)}) - \frac{\sigma^2}{2b^2}(1 - e^{-b(T-t)})^2$$

$$= r_t e^{-b(T-t)} - aC(T - t) - \frac{\sigma^2}{2}C^2(T - t),$$

cf. Relation (5.4), where

$$C(x) = -\frac{1}{b}(1 - e^{-bx}), \qquad x \in \mathbb{R}_+.$$

Forward rate ————

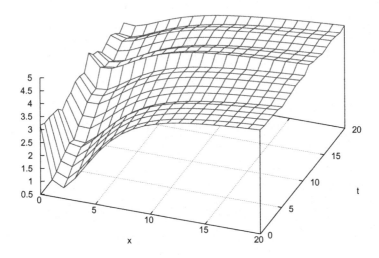

Fig. 6.1   Stochastic process of forward curves.

The short rate process is solution of

$$dr_t = (a - br_t)dt + \sigma dB_t,$$
(6.1)

with

$$r_t = e^{-bt}r_0 + \frac{a}{b}(1 - e^{-bt}) + \sigma \int_0^t e^{-b(t-s)}dB_s$$

$$= f(0,t) + \frac{\sigma^2}{2b^2}(1 - e^{-bt})^2 + \sigma \int_0^t e^{-b(t-s)}dB_s,$$

where

$$f(0,t) = e^{-bt}r_0 + \frac{a}{b}(1 - e^{-bt}) - \frac{\sigma^2}{2b^2}(1 - e^{-bt})^2, \qquad t \in \mathbb{R}_+,$$

is deterministic.

Let us determine the dynamics of the process $(f(t,T))_{t \in [0,T]}$ of forward rates in the Vasicek model. We have

$$d_t f(t, T)$$
$$= e^{-b(T-t)} dr_t + b e^{-b(T-t)} r_t dt + a C'(T-t) dt + \sigma^2 C(T-t) C'(T-t) dt$$
$$= (a - b r_t) e^{-b(T-t)} dt + \sigma e^{-b(T-t)} dB_t + b e^{-b(T-t)} r_t dt$$
$$\quad + a C'(T-t) dt + \sigma^2 C(T-t) C'(T-t) dt$$
$$= -\sigma^2 C(T-t) e^{-b(T-t)} dt + \sigma e^{-b(T-t)} dB_t$$
$$= e^{-b(T-t)} \frac{\sigma^2}{b} (1 - e^{-b(T-t)}) dt + \sigma e^{-b(T-t)} dB_t$$
$$= \sigma^2 e^{-b(T-t)} \int_t^T e^{b(t-s)} ds \, dt + \sigma e^{-b(T-t)} dB_t.$$

Hence $d_t f(t, T)$ can be written as

$$d_t f(t, T) = \alpha(t, T) dt + \sigma(t, T) dB_t$$

with $\sigma(t, T) = \sigma e^{-b(T-t)}$ and

$$\alpha(t, T) = \sigma^2 e^{-b(T-t)} \int_t^T e^{-b(s-t)} ds \qquad (6.2)$$
$$= \sigma(t, T) \int_t^T \sigma(t, s) ds.$$

In the next section we will show that Relation (6.2) between $\alpha(t, T)$ and $\sigma(t, T)$ is not a coincidence, but a general consequence of the absence of arbitrage hypothesis on the dynamics of forward rates. Note also that the coefficient a present in (6.1) has disappeared in (6.2) above. A random simulation of the Vasicek instantaneous forward rates is given in the Figure 6.2.

Recall that for $x = 0$ the first "slice" of this surface is actually the short rate Vasicek process $t \longmapsto r_t = f(t, t) = f(t, t + 0)$ which is represented in Figure 6.3 using another discretization.

Note that at fixed time t, all Vasicek instantaneous forward curves converge to the "long rate"

$$\lim_{x \to \infty} f(t, t + x) = \lim_{T \to \infty} f(t, T) = \frac{a}{b} - \frac{\sigma^2}{2b^2}$$

as x goes to infinity.

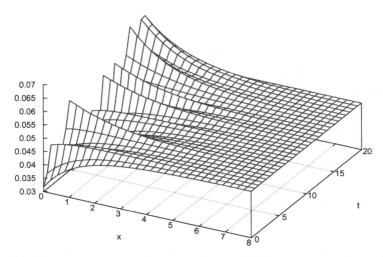

Fig. 6.2   Forward instantaneous curve $(t, x) \longmapsto f(t, t+x)$ in the Vasicek model.

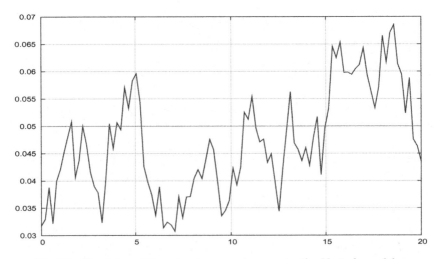

Fig. 6.3   Short term interest rate curve $t \longmapsto r_t$ in the Vasicek model.

In the Musiela notation $(x = T - t)$ we have

$$f(t, T) = f(t, t+x) = r_t e^{-bx} + \frac{a}{b}(1 - e^{-bx}) - \frac{\sigma^2}{2b^2}(1 - e^{-bx})^2$$

$$= \frac{a}{b} - \frac{\sigma^2}{2b^2} + \left(r_t - \frac{a}{b} + \frac{\sigma^2}{b^2}\right)e^{-bx} - \frac{\sigma^2}{2b^2}e^{-2bx},$$

hence for all $t > 0$ the instantaneous forward curves in the Vasicek model "live" in a space of functions generated by

$$x \longmapsto z_1 + z_2 e^{z_3 x} + z_4 e^{z_5 x}, \qquad x \in \mathbb{R}_+, \tag{6.3}$$

with

$$\begin{cases} z_1 = \dfrac{a}{b} - \dfrac{\sigma^2}{2b^2}, \\ z_2 = r_t - \dfrac{a}{b} + \dfrac{\sigma^2}{b^2}, \\ z_3 = -b, \\ z_4 = -\dfrac{\sigma^2}{2b}, \\ z_5 = -2b. \end{cases}$$

Unfortunately, this space of functions is included neither in the Nelson-Siegel space, nor in the Svensson space, of Section 5.4, cf. §3.5 of [Björk (2004)]. A typical realization of such a curve in the Vasicek model is given in Figure 5.4 However it does not appear realistic enough to model the market forward curves considered in the previous chapters, see Figure 6.4.

Fig. 6.4 Market data of LIBOR forward rates $T \longmapsto f(t, T, T + \delta)$.

For this purpose, Svensson curves of the form

$$x \longmapsto g(x) = z_1 + (z_2 + z_3 x)e^{-x z_4} + z_5 x e^{-x z_6}, \qquad x \in \mathbb{R}_+,$$

appear to yield a better modelization, cf. Figure 5.6.

6.3   Spot Forward Rate Dynamics

In the HJM model, the instantaneous forward rate $f(t,T)$ is modeled by a stochastic differential equation of the form

$$d_t f(t,T) = \alpha(t,T)dt + \sigma(t,T)dB_t, \qquad (6.4)$$

where $t \longmapsto \alpha(t,T)$ and $t \longmapsto \sigma(t,T)$, $0 \le t \le T$, are allowed to be random (adapted) processes. In the above equation, the date T is fixed and the differential d_t is with respect to t.

In the Vasicek model the coefficients α and σ are in fact deterministic and we recall that from the previous section we have

$$\alpha(t,T) = \sigma^2 e^{-b(T-t)} \int_t^T e^{-b(T-s)}ds, \qquad \text{and} \qquad \sigma(t,T) = \sigma e^{-b(T-t)}.$$

Next, we determine the dynamics of the spot forward rate

$$f(t,t,T) = \frac{1}{T-t}\int_t^T f(t,s)ds \qquad (6.5)$$

$$= \frac{X_t}{T-t}, \qquad 0 \le t \le T,$$

under the condition (6.4), where

$$X_t = \int_t^T f(t,s)ds = -\log P(t,T), \qquad 0 \le t \le T,$$

and the dynamics of $t \longmapsto f(t,s)$ is given by (6.4). We have

$$d_t X_t = -f(t,t)dt + \int_t^T d_t f(t,s)ds$$

$$= -f(t,t)dt + \int_t^T \alpha(t,s)dsdt + \int_t^T \sigma(t,s)dsdB_t$$

$$= -r_t dt + \left(\int_t^T \alpha(t,s)ds\right)dt + \left(\int_t^T \sigma(t,s)ds\right)dB_t,$$

hence

$$|d_t X_t|^2 = \left(\int_t^T \sigma(t,s)ds\right)^2 dt,$$

and Itô's formula yields, for h a C^2 function,

$$d_t h(t, X_t) = \frac{\partial h}{\partial t}(t, X_t)dt - r_t \frac{\partial h}{\partial x}(t, X_t)dt + \int_t^T \alpha(t, s)ds \frac{\partial h}{\partial x}(t, X_t)dt$$

$$+ \int_t^T \sigma(t, s)ds \frac{\partial h}{\partial x}(t, X_t)dB_t + \frac{1}{2}\left(\int_t^T \sigma(t, s)ds\right)^2 \frac{\partial^2 h}{\partial x^2}(t, X_t)dt.$$

In particular, the dynamics of the spot forward rate (6.5) is

$$d_t f(t, t, T) = \frac{X_t}{(T-t)^2}dt + \frac{1}{T-t}d_t X_t$$

$$= \frac{f(t, t, T)}{T-t}dt - \frac{r_t}{T-t}dt$$

$$+ \frac{1}{T-t}\int_t^T \alpha(t, s)ds dt + \frac{1}{T-t}\int_t^T \sigma(t, s)ds dB_t.$$

In the Vasicek model this gives

$$d_t f(t, t, T) = \frac{f(t, t, T)}{T-t}dt - \frac{r_t}{T-t}dt$$

$$+ \frac{\sigma^2}{T-t}\left(\int_t^T e^{-b(s-t)}\int_t^s e^{-b(t-u)}du ds\right)dt + \frac{\sigma}{T-t}\left(\int_t^T e^{-b(s-t)}ds\right)dB_t.$$

6.4 The HJM Condition

An important question is to determine under which conditions Equation (6.4) makes sense in a financial context, and in particular under which conditions on (6.4) there can be absence of arbitrage.

Under the absence of arbitrage hypothesis, the bond price $P(t, T)$ has been defined (see Chapter 4) as

$$P(t, T) = \mathbb{E}_{\mathbb{Q}}\left[\exp\left(-\int_t^T r_s ds\right)\bigg|\mathcal{F}_t\right]. \tag{6.6}$$

In this framework we have

$$\exp\left(-\int_0^t r_s ds\right)P(t, T) = \exp\left(-\int_0^t r_s ds\right)\mathbb{E}_{\mathbb{Q}}\left[\exp\left(-\int_t^T r_s ds\right)\bigg|\mathcal{F}_t\right]$$

$$= \mathbb{E}_{\mathbb{Q}}\left[\exp\left(-\int_0^t r_s ds\right)\exp\left(-\int_t^T r_s ds\right)\bigg|\mathcal{F}_t\right]$$

$$= \mathbb{E}_{\mathbb{Q}}\left[\exp\left(-\int_0^T r_s ds\right)\bigg|\mathcal{F}_t\right],$$

which is a $\mathbb{Q}$-martingale by the tower property (12.7) of conditional expectations, cf. Appendix A:

$$\mathbb{E}_{\mathbb{Q}}\left[\mathbb{E}_{\mathbb{Q}}\left[\exp\left(-\int_0^T r_s ds\right)\Big|\mathcal{F}_t\right]\Big|\mathcal{F}_u\right] = \mathbb{E}_{\mathbb{Q}}\left[\exp\left(-\int_0^T r_s ds\right)\Big|\mathcal{F}_u\right],$$

$0 < u < t$. Recall also that from Relation (5.1), the instantaneous forward rate $f(t,s)$ satisfies

$$P(t,T) = \exp\left(-\int_t^T f(t,s)ds\right),$$

hence

$$\exp\left(-\int_0^t r_s ds\right)P(t,T) = \exp\left(-\int_0^t r_s ds - \int_t^T f(t,s)ds\right)$$

$$= \exp\left(-\int_0^t r_s ds - X_t\right) \qquad (6.7)$$

is a $\mathbb{Q}$-martingale, $0 \le t \le T$.

Assuming that the short rate $(r_t)_{t\in\mathbb{R}_+}$ has the Markov property the expression (6.6) can be rewritten as

$$P(t,T) = \mathbb{E}_{\mathbb{Q}}\left[\exp\left(-\int_t^T r_s ds\right)\Big|r_t\right]$$

$$= F(t,r_t).$$

Using Itô calculus and the martingale property, the above expression lead us in Chapter 4 to the PDE satisfied by $F(t,x)$.

Here we again apply the same strategy:

(1) Apply Itô's calculus to differentiate (6.7).

(2) Since (6.7) is a $\mathbb{Q}$-martingale under absence of arbitrage, we can equate its dt term to zero.

By Itô's calculus we have

$$d_t e^{-X_t} = -e^{-X_t} d_t X_t + \frac{1}{2}e^{-X_t}(d_t X_t)^2$$

$$= -e^{-X_t} d_t X_t + \frac{1}{2}e^{-X_t}\left(\int_t^T \sigma(t,s)ds\right)^2 dt$$

$$= -e^{-X_t} \left(-r_t dt + \int_t^T \alpha(t,s)dsdt + \int_t^T \sigma(t,s)dsdB_t \right)$$

$$+ \frac{1}{2} e^{-X_t} \left(\int_t^T \sigma(t,s)ds \right)^2 dt,$$

hence

$$d_t \exp\left(-\int_0^t r_s ds - \int_t^T f(t,s)ds \right) = d_t \exp\left(-\int_0^t r_s ds - X_t \right)$$

$$= -r_t \exp\left(-\int_0^t r_s ds - X_t \right) dt$$

$$+ \exp\left(-\int_0^t r_s ds \right) d_t e^{-X_t}$$

$$= -r_t \exp\left(-\int_0^t r_s ds - X_t \right) dt$$

$$- \exp\left(-\int_0^t r_s ds - X_t \right) d_t X_t$$

$$+ \frac{1}{2} \exp\left(-\int_0^t r_s ds - X_t \right) \left(\int_t^T \sigma(t,s)ds \right)^2 dt$$

$$= -r_t \exp\left(-\int_0^t r_s ds - X_t \right) dt$$

$$- \exp\left(-\int_0^t r_s ds - X_t \right) \left(-r_t dt + \int_t^T \alpha(t,s)dsdt + \int_t^T \sigma(t,s)dsdB_t \right)$$

$$+ \frac{1}{2} \exp\left(-\int_0^t r_s ds - X_t \right) \left(\int_t^T \sigma(t,s)ds \right)^2 dt$$

$$= -\exp\left(-\int_0^t r_s ds - X_t \right) \left(\int_t^T \alpha(t,s)dsdt + \int_t^T \sigma(t,s)dsdB_t \right)$$

$$+ \frac{1}{2} \exp\left(-\int_0^t r_s ds - X_t \right) \left(\int_t^T \sigma(t,s)ds \right)^2 dt.$$

Thus the martingale property of the above process implies that

$$\int_t^T \alpha(t,s)ds = \frac{1}{2} \left(\int_t^T \sigma(t,s)ds \right)^2. \tag{6.8}$$

Differentiating the above relation with respect to T, we get

$$\alpha(t,T) = \sigma(t,T) \int_t^T \sigma(t,s)ds, \tag{6.9}$$

which is known as the *HJM absence of arbitrage condition*, cf. [Heath *et al.* (1992)], and is in fact equivalent to (6.8).

As a consequence of Relation (6.9), the stochastic differential equation defining the instantaneous forward rate $f(t,T)$ rewrites as

$$d_t f(t,T) = \sigma(t,T) \left(\int_t^T \sigma(t,s)ds \right) dt + \sigma(t,T)dB_t,$$

and in integral form this gives

$$f(t,T) = f(0,T) + \int_0^t \alpha(s,T)ds + \int_0^t \sigma(s,T)dB_s \tag{6.10}$$

$$= f(0,T) + \int_0^t \sigma(s,T) \int_s^T \sigma(s,u)duds + \int_0^t \sigma(s,T)dB_s.$$

6.5   Markov Property of Short Rates

As noted above, the Markov property of the short rate is of capital importance when deriving the pricing PDE for $P(t,T) = F(t, r_t)$. Thus a natural question is:

- when does the short term interest rate process have the Markov property in the HJM model?

Recall that from Relations (6.9) and (6.10), in the HJM model the short rate process is given by

$$r_t = f(t,t) = f(0,t) + \int_0^t \sigma(s,t) \int_s^t \sigma(s,u)duds + \int_0^t \sigma(s,t)dB_s.$$

In general, a process of the form

$$t \longmapsto Z_t := \int_0^t \sigma(s,t)dB_s, \qquad t \in \mathbb{R}_+, \tag{6.11}$$

where $s \longmapsto \sigma(s,t)$ is $\mathcal{F}_s$-adapted, may *not* be a Markov process due to the dependence of $\sigma(s,t)$ in the variable t.

In fact we have

$$
\begin{aligned}
\mathbb{E}\left[Z_t \middle| \mathcal{F}_u\right] &= \mathbb{E}\left[\int_0^t \sigma(s,t)dB_s \middle| \mathcal{F}_u\right] \\
&= \mathbb{E}\left[\int_0^u \sigma(s,t)dB_s + \int_u^t \sigma(s,t)dB_s \middle| \mathcal{F}_u\right] \\
&= \mathbb{E}\left[\int_0^u \sigma(s,t)dB_s \middle| \mathcal{F}_u\right] + \mathbb{E}\left[\int_u^t \sigma(s,t)dB_s \middle| \mathcal{F}_u\right] \\
&= \mathbb{E}\left[\int_0^u \sigma(s,t)dB_s \middle| \mathcal{F}_u\right] \\
&= \int_0^u \sigma(s,t)dB_s,
\end{aligned}
\tag{6.12}
$$

where we applied Relation (1.6) of Chapter 1.

According to the Markov property the above quantity should depend only on u, t, and

$$
Z_u = \int_0^u \sigma(s,u)dB_s,
$$

and there is a priori no reason for this property to hold here, as follows by inspection of (6.12).

Nevertheless, the Markov property of Z_t defined as in (6.11) does hold for some particular choices of $\sigma(t,T)$. For example, in case

$$
\sigma(s,t) = e^{-b(t-s)},
$$

we have

$$
\begin{aligned}
\mathbb{E}\left[Z_t \middle| \mathcal{F}_u\right] &= \int_0^u \sigma(s,t)dB_s \\
&= \int_0^u e^{-b(t-s)}dB_s \\
&= e^{-b(t-u)}\int_0^u e^{-b(u-s)}dB_s \\
&= e^{-b(t-u)}Z_u,
\end{aligned}
$$

as known from the explicit solution to the Vasicek model, cf. (4.8).

More generally, the Markov property does hold for a stochastic integral process of the form

$$
t \longmapsto \int_0^t \sigma(s,t)dB_s,
$$

under the product condition

$$\sigma(s,t) = \xi(s)\psi(t), \qquad 0 \le s \le t, \tag{6.13}$$

where $\xi(s)$ and $\psi(t)$ are two deterministic functions. Indeed we have

$$\begin{aligned}
\mathbb{E}\left[Z_t \middle| \mathcal{F}_u\right] &= \int_0^u \sigma(s,t)dB_s \\
&= \psi(t)\int_0^u \xi(s)dB_s \\
&= \frac{\psi(t)}{\psi(u)}\int_0^u \psi(u)\xi(s)dB_s \\
&= \frac{\psi(t)}{\psi(u)}Z_u, \qquad 0 \le u \le t.
\end{aligned}$$

Recall that in the Vasicek model we have $\sigma(s,t) = \sigma e^{-b(t-s)}$, $0 \le s \le t$, hence condition (6.13) holds and the short rate is indeed a Markov process.

6.6   The Hull-White Model

Our goal is now to derive a stochastic differential equation satisfied by the short rate process $(r_t)_{t\in\mathbb{R}_+}$ in the Markovian HJM model under the product assumption (6.13) on the volatility coefficient $\sigma(s,t)$. In this way we will recover the time-dependent Hull-White short rate model described in Section 3.3.

By (6.4) and the HJM Condition (6.9), or directly by (6.10), we have

$$\begin{aligned}
r_t &= f(t,t) \\
&= f(0,t) + \int_0^t \sigma(s,t)\int_s^t \sigma(s,u)duds + \int_0^t \sigma(s,t)dB_s \\
&= f(0,t) + \int_0^t \xi(s)\psi(t)\int_s^t \xi(s)\psi(u)duds + \psi(t)\int_0^t \xi(s)dB_s,
\end{aligned}$$

hence

$$r_t = U(t) + \psi(t)\int_0^t \xi(s)dB_s, \tag{6.14}$$

where

$$U(t) = f(0,t) + \psi(t)\int_0^t \xi^2(s)\int_s^t \psi(u)duds.$$

Using the relation

$$\int_0^t \xi(s)dB_s = \frac{r_t - U(t)}{\psi(t)}$$

that follows from (6.14), we have

$$dr_t = U'(t)dt + \psi'(t)\left(\int_0^t \xi(s)dB_s\right)dt + \psi(t)\xi(t)dB_t$$

$$= U'(t)dt + (r_t - U(t))\frac{\psi'(t)}{\psi(t)}dt + \psi(t)\xi(t)dB_t,$$

which indeed shows that the short rate process $(r_t)_{t \in \mathbb{R}_+}$ has the Markov property as the solution of a stochastic differential equation, cf. Property 4.1.

The above equation belongs to the class of [Hull and White (1990)] short rate models of the form

$$dr_t = (a(t) - b(t)r_t)dt + \sigma(t)dB_t,$$

cf. Section 3.3, which can be interpreted as time-dependent Vasicek models with explicit solution

$$r_t = r_s e^{-\int_s^t b(\tau)d\tau} + \int_s^t e^{-\int_u^t b(\tau)d\tau} a(u)du + \int_s^t \sigma(u)e^{-\int_u^t b(\tau)d\tau}dB_u,$$

$0 \le s \le t$.

6.7 Exercises

Exercise 6.1. (Exercise 5.1 continued).

(1) Derive the stochastic equation satisfied by the instantaneous forward rate $f(t, T)$.
(2) Check that the HJM absence of arbitrage Condition (6.9) is satisfied in this equation.

Exercise 6.2. (Exercise 5.2 continued).

(1) Derive the stochastic equation satisfied by the instantaneous forward rate $f(t, T)$.
(2) Check that the HJM absence of arbitrage condition is satisfied in the equation of Question (1).

Chapter 7

The Forward Measure and Derivative Pricing

In this chapter we introduce the notion of forward measure for the pricing of interest rate derivatives. We use the Girsanov theorem to obtain the dynamics of the short rate processes under forward measures, with explicit calculations in the case of the Vasicek model.

7.1 Forward Measure

In a standard Black-Scholes framework with a riskless account yielding interests at the instantaneous short rate r_t, the price at time t of a contingent claim with payoff F at exercise time T is computed as the conditional expectation

$$\mathbb{E}_{\mathbb{Q}}\left[e^{-\int_t^T r_s ds} F \middle| \mathcal{F}_t\right]$$

under a risk neutral probability measure $\mathbb{Q}$, cf. (2.16). When the interest rate process $(r_t)_{t \in \mathbb{R}_+}$ is a deterministic function of time, this expression becomes

$$e^{-\int_t^T r_s ds} \mathbb{E}_{\mathbb{Q}}[F \mid \mathcal{F}_t],$$

and when $(r_t)_{t \in \mathbb{R}_+}$ equals a deterministic constant r we get the well-known expression

$$e^{-(T-t)r} \mathbb{E}_{\mathbb{Q}}[F \mid \mathcal{F}_t].$$

In most interest rate models the short term interest rate $(r_t)_{t \in \mathbb{R}_+}$ is a random process and the above manipulation will not be allowed, meaning we will have to evaluate expressions of the form

$$\mathbb{E}_{\mathbb{Q}}\left[e^{-\int_t^T r_s ds} F \middle| \mathcal{F}_t\right] \tag{7.1}$$

where $(r_t)_{t\in\mathbb{R}_+}$ will be a random process, adding another level of complexity in comparison with the standard Black-Scholes framework of Chapter 2.

Note that when computing the prices of bonds in Chapter 4 we have already evaluated such expressions as the solution of a PDE in the case of a constant payoff $F = \$1$. In case payoff F is random and of the form $F = h(f(T,T,S))$ - e.g. for an option on the spot forward rate $f(T,T,S)$ - the computation of (7.1) would require the knowledge of the joint distribution of $\int_t^T r_s ds$ and $f(T,T,S)$, which can lead to complex computations.

Following the choice made in Section 4.2 we take $\mathbb{Q} = \mathbb{P}$ and we work under the assumption of absence of arbitrage under $\mathbb{P}$, which states that

$$t \longmapsto e^{-\int_0^t r_s ds} P(t,T), \qquad 0 \le t \le T, \tag{7.2}$$

is an $\mathcal{F}_t$-martingale under $\mathbb{P}$.

Definition 7.1. *The forward measure with maturity T is the probability measure $\tilde{\mathbb{P}}$ defined as*

$$\frac{d\tilde{\mathbb{P}}}{d\mathbb{P}} = \frac{1}{P(0,T)} e^{-\int_0^T r_s ds}.$$

In the sequel, the expectation under $\tilde{\mathbb{P}}$ will be denoted by $\mathbb{E}_{\tilde{\mathbb{P}}}$.

The following proposition will allow us to price contingent claims using the forward measure $\tilde{\mathbb{P}}$.

Proposition 7.1. *For all sufficiently integrable random variables F we have*

$$\mathbb{E}_{\mathbb{P}}\left[Fe^{-\int_t^T r_s ds}\Big|\mathcal{F}_t\right] = P(t,T)\,\mathbb{E}_{\tilde{\mathbb{P}}}[F \mid \mathcal{F}_t], \qquad 0 \le t \le T. \tag{7.3}$$

Proof. Indeed[1], for all bounded and $\mathcal{F}_t$-measurable random variables G we have

$$\mathbb{E}_{\mathbb{P}}\left[GFe^{-\int_t^T r_s ds}\right] = P(0,T)\,\mathbb{E}_{\tilde{\mathbb{P}}}\left[Ge^{\int_0^t r_s ds}F\right]$$

$$= P(0,T)\,\mathbb{E}_{\tilde{\mathbb{P}}}\left[Ge^{\int_0^t r_s ds}\,\mathbb{E}_{\tilde{\mathbb{P}}}[F \mid \mathcal{F}_t]\right]$$

$$= P(0,T)\,\mathbb{E}_{\mathbb{P}}\left[\frac{d\tilde{\mathbb{P}}}{d\mathbb{P}}Ge^{\int_0^t r_s ds}\,\mathbb{E}_{\tilde{\mathbb{P}}}[F \mid \mathcal{F}_t]\right]$$

[1]We use the characterization $X = \mathbb{E}[F|\mathcal{F}_t] \iff \mathbb{E}[GX] = \mathbb{E}[GF]$ for all G bounded and $\mathcal{F}_t$-measurable, cf. Relation (12.4) in Appendix A.

$$= P(0,T)\, \mathbb{E}_{\mathbb{P}} \left[\frac{1}{P(0,T)} e^{-\int_0^T r_s ds} G e^{\int_0^t r_s ds}\, \mathbb{E}_{\tilde{\mathbb{P}}}[F \mid \mathcal{F}_t] \right]$$

$$= \mathbb{E}_{\mathbb{P}} \left[G e^{-\int_t^T r_s ds}\, \mathbb{E}_{\tilde{\mathbb{P}}}[F \mid \mathcal{F}_t] \right]$$

$$= \mathbb{E}_{\mathbb{P}} \left[G P(t,T)\, \mathbb{E}_{\tilde{\mathbb{P}}}[F \mid \mathcal{F}_t] \right],$$

where on the last line we used the above characterization of the conditional expectation and the facts that we have

$$P(t,T) = \mathbb{E}_{\mathbb{P}} \left[e^{-\int_t^T r_s ds} \middle| \mathcal{F}_t \right], \tag{7.4}$$

on the one hand, and that G, $\mathbb{E}_{\tilde{\mathbb{P}}}[F \mid \mathcal{F}_t]$ are $\mathcal{F}_t$-measurable random variables on the other hand. $\qquad\square$

As a consequence of this proposition, the computation of $\mathbb{E}_{\mathbb{P}} \left[F e^{-\int_t^T r_s ds} \middle| \mathcal{F}_t \right]$ can be replaced by that of $P(t,T)\, \mathbb{E}_{\tilde{\mathbb{P}}}[F \mid \mathcal{F}_t]$ under the *forward* measure $\tilde{\mathbb{P}}$.

As a corollary of Proposition 7.1, the next lemma tells us how the density $d\tilde{\mathbb{P}}/d\mathbb{P}$ behaves under conditioning with respect to $\mathcal{F}_t$. Recall that by definition,

$$\Lambda_t := \frac{d\tilde{\mathbb{P}}_{|\mathcal{F}_t}}{d\mathbb{P}_{|\mathcal{F}_t}}, \qquad 0 \le t \le T,$$

is the only random variable to satisfy

$$\mathbb{E}_{\tilde{\mathbb{P}}}[F \mid \mathcal{F}_t] = \mathbb{E}_{\mathbb{P}} \left[F \Lambda_t \middle| \mathcal{F}_t \right],$$

i.e.

$$\int_\Omega F d\tilde{\mathbb{P}}_{|\mathcal{F}_t} = \int_\Omega F \Lambda_t d\mathbb{P}_{|\mathcal{F}_t},$$

for all bounded random variables F.

Lemma 7.1. *We have*

$$\frac{d\tilde{\mathbb{P}}_{|\mathcal{F}_t}}{d\mathbb{P}_{|\mathcal{F}_t}} = \frac{e^{-\int_t^T r_s ds}}{P(t,T)}, \qquad 0 \le t \le T. \tag{7.5}$$

Proof. Rewrite (7.3) as

$$\mathbb{E}_{\tilde{\mathbb{P}}}[F \mid \mathcal{F}_t] = \mathbb{E}_{\mathbb{P}} \left[F \frac{e^{-\int_t^T r_s ds}}{P(t,T)} \middle| \mathcal{F}_t \right], \qquad 0 \le t \le T,$$

for all F bounded and measurable, which implies (7.5). $\qquad\square$

Note that $\dfrac{d\tilde{\mathbb{P}}_{|\mathcal{F}_t}}{d\mathbb{P}_{|\mathcal{F}_t}}$ is *not* equal to $\mathbb{E}_{\mathbb{P}}\left[\dfrac{d\tilde{\mathbb{P}}}{d\mathbb{P}}\middle|\mathcal{F}_t\right]$, in fact we have

$$\mathbb{E}_{\mathbb{P}}\left[\frac{d\tilde{\mathbb{P}}}{d\mathbb{P}}\middle|\mathcal{F}_t\right] = \frac{1}{P(0,T)}\mathbb{E}_{\mathbb{P}}\left[e^{-\int_0^T r_s ds}\middle|\mathcal{F}_t\right] \qquad (7.6)$$

$$= \frac{P(t,T)}{P(0,T)}e^{-\int_0^t r_s ds}, \qquad 0 \le t \le T,$$

by the assumption (7.2).

In addition we have the following result.

Proposition 7.2. *For all $S, T \ge 0$, the process*

$$t \longmapsto \frac{P(t,S)}{P(t,T)}, \qquad 0 \le t \le S \wedge T,$$

is an $\mathcal{F}_t$-martingale under $\tilde{\mathbb{P}}$, provided it is integrable.

Proof.　For all bounded and $\mathcal{F}_s$-measurable random variables F, from Relation (7.6) we have:[2]

$$\mathbb{E}_{\tilde{\mathbb{P}}}\left[F\frac{P(t,S)}{P(t,T)}\right] = \mathbb{E}_{\mathbb{P}}\left[F\frac{e^{-\int_0^T r_u du}}{P(0,T)}\frac{P(t,S)}{P(t,T)}\right]$$

$$= \frac{1}{P(0,T)}\mathbb{E}_{\mathbb{P}}\left[Fe^{-\int_0^t r_u du}P(t,S)\right]$$

$$= \frac{1}{P(0,T)}\mathbb{E}_{\mathbb{P}}\left[Fe^{-\int_0^s r_u du}P(s,S)\right]$$

$$= \mathbb{E}_{\mathbb{P}}\left[F\frac{e^{-\int_0^T r_u du}}{P(0,T)}\frac{P(s,S)}{P(s,T)}\right]$$

$$= \mathbb{E}_{\tilde{\mathbb{P}}}\left[F\frac{P(s,S)}{P(s,T)}\right],$$

hence

$$\mathbb{E}_{\tilde{\mathbb{P}}}\left[\frac{P(t,S)}{P(t,T)}\middle|\mathcal{F}_s\right] = \frac{P(s,S)}{P(s,T)}. \qquad \square$$

Note that Proposition 7.2 can also be proved by Lemma 7.1, as follows:

$$\mathbb{E}_{\tilde{\mathbb{P}}}\left[\frac{P(t,S)}{P(t,T)}\middle|\mathcal{F}_s\right] = \mathbb{E}_{\mathbb{P}}\left[\frac{P(t,S)}{P(t,T)}\frac{d\tilde{\mathbb{P}}_{|\mathcal{F}_s}}{d\mathbb{P}_{|\mathcal{F}_s}}\middle|\mathcal{F}_s\right]$$

---

[2]We use the characterization $X = \mathbb{E}[F|\mathcal{F}_t] \iff \mathbb{E}[GX] = \mathbb{E}[GF]$ for all G bounded and $\mathcal{F}_t$-measurable, cf. Relation (12.4) in Appendix A.

$$= \mathbb{E}_{\mathbb{P}} \left[\frac{P(t,S)}{P(t,T)} \frac{e^{- \int_s^T r_u du}}{P(s,T)} \middle| \mathcal{F}_s \right]$$

$$= \frac{1}{P(s,T)} \mathbb{E}_{\mathbb{P}} \left[P(t,S) \frac{e^{- \int_s^T r_u du}}{P(t,T)} \middle| \mathcal{F}_s \right]$$

$$= \frac{1}{P(s,T)} \mathbb{E}_{\mathbb{P}} \left[\mathbb{E}_{\mathbb{P}} \left[\frac{P(t,S)}{P(t,T)} e^{- \int_s^T r_u du} \middle| \mathcal{F}_t \right] \middle| \mathcal{F}_s \right]$$

$$= \frac{1}{P(s,T)} \mathbb{E}_{\mathbb{P}} \left[e^{- \int_s^t r_u du} \frac{P(t,S)}{P(t,T)} \mathbb{E}_{\mathbb{P}} \left[e^{- \int_t^T r_u du} \middle| \mathcal{F}_t \right] \middle| \mathcal{F}_s \right]$$

$$= \frac{1}{P(s,T)} \mathbb{E}_{\mathbb{P}} \left[e^{- \int_s^t r_u du} P(t,S) \middle| \mathcal{F}_s \right]$$

$$= \frac{e^{\int_0^s r_u du}}{P(s,T)} \mathbb{E}_{\mathbb{P}} \left[e^{- \int_0^t r_u du} P(t,S) \middle| \mathcal{F}_s \right]$$

$$= \frac{P(s,S)}{P(s,T)},$$

$0 \leq s \leq t \leq T \leq S$, since the discounted bond price process $t \longmapsto$ $e^{- \int_0^t r_u du} P(t,S)$ is a martingale under $\mathbb{P}$.

7.2 Dynamics under the Forward Measure

In order to apply Proposition 7.1 and to compute the price $\mathbb{E}_{\mathbb{P}} \left[e^{- \int_t^T r_s ds} F \middle| \mathcal{F}_t \right]$ from the evaluation of

$$P(t,T) \, \mathbb{E}_{\tilde{\mathbb{P}}}[F \mid \mathcal{F}_t],$$

we will need to determine the dynamics of the underlying processes r_t, $f(t,T,S)$, and $P(t,T)$ under the forward measure $\tilde{\mathbb{P}}$.

For this we will assume that the dynamics of $P(t,T)$ under $\mathbb{P}$ has the form

$$\frac{dP(t,T)}{P(t,T)} = r_t dt + \zeta_t dB_t, \qquad 0 \leq t \leq T, \tag{7.7}$$

where $(B_t)_{t \in \mathbb{R}_+}$ is a standard Brownian motion under $\mathbb{P}$ and $(r_t)_{t \in \mathbb{R}_+}$ and $(\zeta(t))_{t \in \mathbb{R}_+}$ are adapted processes with respect to the filtration $(\mathcal{F}_t)_{t \in \mathbb{R}_+}$ generated by $(B_t)_{t \in \mathbb{R}_+}$.

An application of Itô's calculus to (7.7) shows that

$$d \left(e^{- \int_0^t r_s ds} P(t,T) \right) = \zeta_t \left(e^{- \int_0^t r_s ds} P(t,T) \right) dB_t, \tag{7.8}$$

which is consistent with the fact that

$$t \longmapsto e^{-\int_0^t r_s ds} P(t, T) \tag{7.9}$$

is a martingale under $\mathbb{P}$, cf. Proposition 7.2 and Corollary 1.1.

In order to determine the dynamics of the underlying processes under $\tilde{\mathbb{P}}$ we will use the following proposition which is obtained from the Girsanov Theorem 2.1.

Proposition 7.3. *The process*

$$\tilde{B}_t := B_t - \int_0^t \zeta_s ds, \qquad 0 \le t \le T, \tag{7.10}$$

is a standard Brownian motion under $\tilde{\mathbb{P}}$.

Proof.   Letting

$$\Psi(t) = \mathbb{E}_{\mathbb{P}} \left[\frac{d\tilde{\mathbb{P}}}{d\mathbb{P}} \bigg| \mathcal{F}_t \right]$$

$$= \frac{1}{P(0, T)} \mathbb{E}_{\mathbb{P}} \left[e^{-\int_0^T r_s ds} \bigg| \mathcal{F}_t \right]$$

$$= \frac{P(t, T)}{P(0, T)} e^{-\int_0^t r_s ds}, \qquad 0 \le t \le T.$$

Equation (7.8) rewrites as

$$d\Psi(t) = \Psi(t) \zeta_t dB_t,$$

which is solved as

$$\Psi(t) = \exp\left(\int_0^t \zeta_s dB_s - \frac{1}{2} \int_0^t |\zeta_s|^2 ds \right),$$

hence

$$\mathbb{E}_{\mathbb{P}} \left[\frac{d\tilde{\mathbb{P}}}{d\mathbb{P}} \bigg| \mathcal{F}_T \right] = \Psi(T) = \exp\left(\int_0^T \zeta_s dB_s - \frac{1}{2} \int_0^T |\zeta_s|^2 ds \right),$$

and we conclude by the Girsanov Theorem 2.1.   □

As a consequence of Proposition 7.3, the dynamics of $t \longmapsto P(t, T)$ under $\tilde{\mathbb{P}}$ is now given by

$$\frac{dP(t, T)}{P(t, T)} = r_t dt + |\zeta_t|^2 dt + \zeta_t d\tilde{B}_t,$$

where $(\tilde{B}_t)_{t\in\mathbb{R}_+}$ is a standard Brownian motion under $\tilde{\mathbb{P}}$, and we also have

$$d\left(e^{-\int_0^t r_s ds}P(t,T)\right) = |\zeta_t|^2 e^{-\int_0^t r_s ds}P(t,T)dt + \zeta_t e^{-\int_0^t r_s ds}P(t,T)d\tilde{B}_t.$$

In the case of two bonds with maturities T and S, with prices $P(t,T)$ and $P(t,S)$ given by

$$\frac{dP(t,T)}{P(t,T)} = r_t dt + \zeta_t^T dB_t,$$

and

$$\frac{dP(t,S)}{P(t,S)} = r_t dt + \zeta_t^S dB_t,$$

Itô's formula yields

$$d\left(\frac{P(t,S)}{P(t,T)}\right) = \frac{P(t,S)}{P(t,T)}(\zeta^S(t) - \zeta^T(t))(dB_t - \zeta^T(t)dt)$$

$$= \frac{P(t,S)}{P(t,T)}(\zeta^S(t) - \zeta^T(t))d\tilde{B}_t, \tag{7.11}$$

where $(\tilde{B}_t)_{t\in\mathbb{R}_+}$ is a standard Brownian motion under $\tilde{\mathbb{P}}$, which, from Proposition 7.10, recovers the martingale property of $P(t,S)/P(t,T)$ stated in Proposition 7.2.

In case the short rate process $(r_t)_{t\in\mathbb{R}_+}$ is Markovian and solution of

$$dr_t = \mu(t,r_t)dt + \sigma(t,r_t)dB_t,$$

its dynamics will be given under $\tilde{\mathbb{P}}$ by

$$dr_t = \mu(t,r_t)dt + \sigma(t,r_t)\zeta_t dt + \sigma(t,r_t)d\tilde{B}_t.$$

Recall that in this Markovian setting, the bond price $P(t,T)$ is expressed as

$$P(t,T) = \mathbb{E}_{\mathbb{P}}\left[Fe^{-\int_t^T r_s ds}\Big|\mathcal{F}_t\right]$$

$$= \mathbb{E}_{\mathbb{P}}\left[Fe^{-\int_t^T r_s ds}\Big|r_t\right]$$

$$= F(t,r_t),$$

i.e. it becomes a function $F(t,r_t)$ of t and r_t. Itô's formula then shows that

$$d\left(e^{-\int_0^t r_s ds}P(t,T)\right) = e^{-\int_0^t r_s ds}\sigma(t,r_t)\frac{\partial F}{\partial x}(t,r_t)dB_t,$$

since, by Corollary II-1 of [Protter (2005)]), the sum of all terms in dt vanish in the above expression because

$$t \longmapsto e^{-\int_0^t r_s ds}P(t,T) = e^{-\int_0^t r_s ds}F(t,r_t),$$

is a martingale under $\mathbb{P}$ from (7.9) or Proposition 7.2.

Hence the dynamics of $(P(t,T))_{t \in [0,T]}$ is given by

$$\frac{dP(t,T)}{P(t,T)} = r_t dt + \sigma(t,r_t) \frac{1}{P(t,T)} \frac{\partial F}{\partial x}(t,r_t) dB_t$$

$$= r_t dt + \sigma(t,r_t) \frac{1}{F(t,r_t)} \frac{\partial F}{\partial x}(t,r_t) dB_t$$

$$= r_t dt + \sigma(t,r_t) \frac{\partial \log F}{\partial x}(t,r_t) dB_t,$$

and the process $(\zeta_t)_{t \in \mathbb{R}_+}$ in (7.7) is given by

$$\zeta_t = \sigma(t,r_t) \frac{\partial \log F}{\partial x}(t,r_t), \qquad 0 \le t \le T.$$

As an example, in the Vasicek model, where $\sigma(t,x)$ is constant equal to σ, the price $P(t,T)$ has the form

$$P(t,T) = F(t,r_t) = e^{C(T-t)r_t + A(T-t)},$$

where

$$C(T-t) = -\frac{1}{b}(1 - e^{-b(T-t)}),$$

hence

$$\log F(t,r_t) = C(T-t)r_t + A(T-t),$$

and

$$\zeta_t = \sigma C(T-t) = -\frac{\sigma}{b}(1 - e^{-b(T-t)}).$$

7.3   Derivative Pricing

Using the above framework we are now able to compute the price at time t of a contingent claim with random payoff F at exercise time T from the equality

$$\mathbb{E}_{\mathbb{P}}\left[e^{-\int_t^T r_s ds} F \,\Big|\, \mathcal{F}_t\right] = P(t,T)\, \mathbb{E}_{\tilde{\mathbb{P}}}\left[F \,\Big|\, \mathcal{F}_t\right]$$

and the knowledge of the dynamics of r_t under the probability measure $\tilde{\mathbb{P}}$.

Let us consider a bond call option on $P(T,S)$ with payoff

$$F = (P(T,S) - K)^+$$

and price

$$\mathbb{E}_{\mathbb{P}}\left[e^{-\int_t^T r_s ds}(P(T,S)-K)^+\Big|\mathcal{F}_t\right] = P(t,T)\,\mathbb{E}_{\tilde{\mathbb{P}}}\left[(P(T,S)-K)^+\Big|\mathcal{F}_t\right]$$

in the case where $P(t,T)$ is given for all $T>0$ by

$$\frac{dP(t,T)}{P(t,T)} = r_t dt + \zeta_t^T dB_t, \qquad 0 \le t \le T,$$

where the process $(\zeta_t^T)_{t\in[0,T]}$ is adapted. From (7.11) we have

$$P(T,S) = \frac{P(t,S)}{P(t,T)}\exp\left(\int_t^T(\zeta^S(s)-\zeta^T(s))d\tilde{B}_s - \frac{1}{2}\int_t^T|\zeta^S(s)-\zeta^T(s)|^2 ds\right),$$

and when $(\zeta^T(s))_{s\in[0,T]}$, $(\zeta^S(s))_{s\in[0,S]}$ are deterministic, we get

$$\mathbb{E}_{\mathbb{P}}\left[e^{-\int_t^T r_s ds}(P(T,S)-K)^+\Big|\mathcal{F}_t\right]$$

$$= P(t,T)\,\mathbb{E}_{\tilde{\mathbb{P}}}\left[\left(\frac{P(t,S)}{P(t,T)}\exp\left(X-\frac{1}{2}\int_t^T|\zeta^S(s)-\zeta^T(s)|^2 ds\right)-K\right)^+\Big|\mathcal{F}_t\right]$$

$$= P(t,T)\,\mathbb{E}_{\tilde{\mathbb{P}}}\left[\left(e^{X+m(t,T,S)}-K\right)^+\Big|\mathcal{F}_t\right],$$

where X is a centered Gaussian random variable with mean

$$m(t,T,S) = -\frac{1}{2}v^2(t,T,S) + \log\frac{P(t,S)}{P(t,T)},$$

and variance

$$v^2(t,T,S) = \int_t^T|\zeta^S(s)-\zeta^T(s)|^2 ds,$$

given $\mathcal{F}_t$. Recall that by Lemma 2.3, when X is a centered Gaussian random variable with variance v^2, the expectation of $(e^{m+X}-K)^+$ is given, as in the standard Black-Scholes formula, by

$$\mathbb{E}[(e^{m+X}-K)^+] = e^{m+\frac{v^2}{2}}\Phi(v+(m-\log K)/v) - K\Phi((m-\log K)/v),$$

where

$$\Phi(z) = \int_{-\infty}^z e^{-y^2/2}\frac{dy}{\sqrt{2\pi}}, \qquad z \in \mathbb{R},$$

denotes the Gaussian cumulative distribution function and for simplicity of notation we dropped the indices t,T,S in $m(t,T,S)$ and $v^2(t,T,S)$.

Consequently we have

$$\mathbb{E}_{\mathbb{P}}\left[e^{-\int_t^T r_s ds}(P(T,S)-K)^+\Big|\mathcal{F}_t\right] \qquad\qquad (7.12)$$

$$= P(t,S)\Phi\left(\frac{v}{2} + \frac{1}{v}\log\frac{P(t,S)}{KP(t,T)}\right) - KP(t,T)\Phi\left(-\frac{v}{2} + \frac{1}{v}\log\frac{P(t,S)}{KP(t,T)}\right).$$

In the Vasicek model we have

$$dr_t = (a - br_t)dt + \sigma dB_t,$$

and

$$\zeta_t = \sigma C(T - t) = -\frac{\sigma}{b}(1 - e^{-b(T-t)}),$$

hence v is given by

$$v^2 = v^2(t, T, S) = \frac{\sigma^2}{b^2}\int_t^T (e^{-b(T-s)} - e^{-b(S-s)})^2 ds.$$

In the remaining of this section we show how (7.12) can be recovered, still in the Vasicek model, by an independent computation based on the dynamics of r_t and $P(t,T)$ under $\tilde{\mathbb{P}}$, respectively given by

$$dr_t = (a - br_t)dt - \frac{\sigma^2}{b}(1 - e^{-b(T-t)})dt + \sigma d\tilde{B}_t \qquad (7.13)$$

and

$$\frac{dP(t,T)}{P(t,T)} = r_t dt + \frac{\sigma^2}{b^2}(1 - e^{-b(T-t)})^2 dt - \frac{\sigma}{b}(1 - e^{-b(T-t)})d\tilde{B}_t.$$

Equation (7.13) can be solved as

$$r_t = r_s e^{-b(t-s)} + \int_s^t e^{-b(t-u)}(a + \sigma^2 C(T - u))du + \sigma\int_s^t e^{-b(t-u)}d\tilde{B}_u$$

$$= \mathbb{E}_{\tilde{\mathbb{P}}}[r_t \mid \mathcal{F}_s] + \sigma\int_s^t e^{-b(t-u)}d\tilde{B}_u,$$

hence from Proposition 1.2, the conditional mean and variance of r_t under $\tilde{\mathbb{P}}$ are

$$\mathbb{E}_{\tilde{\mathbb{P}}}[r_t \mid \mathcal{F}_s] = r_s e^{-b(t-s)} + \int_s^t e^{-b(t-u)}(a + \sigma^2 C(T - u))du$$

and

$$\mathrm{Var}_{\tilde{\mathbb{P}}}[r_t \mid \mathcal{F}_s] = \mathbb{E}_{\tilde{\mathbb{P}}}[(r_t - \mathbb{E}_{\tilde{\mathbb{P}}}[r_t \mid \mathcal{F}_s])^2 \mid \mathcal{F}_s]$$

$$= \frac{\sigma^2}{2b}(1 - e^{-2b(t-s)}).$$

Hence the price the bond call option on $P(T,S)$ is

$$\mathbb{E}_{\mathbb{P}}\left[e^{-\int_t^T r_s ds}(P(T,S) - K)^+ \Big| \mathcal{F}_t\right] = P(t,T)\mathbb{E}_{\tilde{\mathbb{P}}}\left[(P(T,S) - K)^+ \Big| \mathcal{F}_t\right]$$

$$= P(t,T)\,\mathbb{E}_{\tilde{\mathbb{P}}}\left[(F(T,r_T)-K)^+\Big|\mathcal{F}_t\right]$$

$$= P(t,T)\,\mathbb{E}_{\tilde{\mathbb{P}}}\left[(e^{A(S-T)+r_T C(S-T)}-K)^+\Big|\mathcal{F}_t\right]$$

$$= P(t,T)\,\mathbb{E}_{\tilde{\mathbb{P}}}\left[(e^{m(t,T,S)+X}-K)^+\Big|r_t\right]$$

where X is a centered Gaussian random variable with variance

$$\begin{aligned} v^2(t,T,S) &= \mathrm{Var}_{\tilde{\mathbb{P}}}[C(S-T)r_T \mid \mathcal{F}_t]\\ &= C^2(S-T)\,\mathrm{Var}_{\tilde{\mathbb{P}}}[r_T \mid r_t]\\ &= \frac{\sigma^2}{2b}C^2(S-T)(1-e^{-2b(T-t)}) \end{aligned} \tag{7.14}$$

given $\mathcal{F}_t$, and

$$\tilde{m}(t,T,S) = A(S-T) + C(S-T)\,\mathbb{E}_{\mathbb{P}}[r_T \mid \mathcal{F}_t]$$

$$= A(S-T) + C(S-T)\left(r_t e^{-b(T-t)} + \int_t^T e^{-b(T-u)}(a+\sigma C(T-u))du\right),$$

where

$$A(S-T) = \frac{4ab-3\sigma^2}{4b^3} + \frac{\sigma^2-2ab}{2b^2}(S-T) + \frac{\sigma^2-ab}{b^3}e^{-b(S-T)} - \frac{\sigma^2}{4b^3}e^{-2b(S-T)}.$$

Since from Proposition 7.2,

$$t \longmapsto \frac{P(t,S)}{P(t,T)}, \qquad 0 \le t \le T \le S,$$

is a martingale under $\tilde{\mathbb{P}}$, we have the relation

$$\begin{aligned} \frac{P(t,S)}{P(t,T)} &= \mathbb{E}_{\tilde{\mathbb{P}}}\left[P(T,S)\Big|\mathcal{F}_t\right]\\ &= \mathbb{E}_{\tilde{\mathbb{P}}}\left[e^{A(S-T)+r_T C(S-T)}\Big|\mathcal{F}_t\right]\\ &= e^{A(S-T)+C(S-T)\,\mathbb{E}_{\mathbb{P}}[r_T|\mathcal{F}_t]+\frac{1}{2}C^2(S-T)\,\mathrm{Var}_{\tilde{\mathbb{P}}}[r_T|\mathcal{F}_t]}\\ &= e^{\tilde{m}(t,T,S)+\frac{1}{2}v^2(t,T,S)}. \end{aligned} \tag{7.15}$$

The above *non-trivial* Relation (7.15) can actually be checked by hand[3]:

$$\begin{aligned} -v^2/2 + \log(P(t,S)/P(t,T)) &= -v^2/2 + \log P(t,S) - \log P(t,T)\\ &= -v^2/2 + A(S-t) + r_t C(S-t) - (A(T-t) + r_t C(T-t))\\ &= -v^2/2 + A(S-t) - A(T-t) + r_t(C(S-t) - C(T-t))\\ &= -\frac{\sigma^2}{4b}C^2(S-T)(1-e^{-2b(T-t)}) \end{aligned}$$

---

[3]This can take a significant amount of time.

$$+A(S - t) - A(T - t) + r_t C(S - T)e^{-b(T-t)}$$
$$= A(S - T)$$
$$+C(S - T)\left(r_t e^{-b(T-t)} + \int_t^T e^{-b(T-u)}(a + \sigma^2 C(T - u))du\right)$$
$$= \tilde{m}(t, T, S).$$

Using (7.14) and (7.15) we finally recover (7.12), i.e.

$$\mathbb{E}_{\mathbb{P}}\left[e^{-\int_t^T r_s ds}(P(T, S) - K)^+ \Big| \mathcal{F}_t\right] = P(t, T)\,\mathbb{E}_{\tilde{\mathbb{P}}}\left[(e^{\tilde{m}(t, T, S) + X} - K)^+ \Big| r_t\right]$$
$$= P(t, T)e^{\tilde{m} + \frac{v^2}{2}}\Phi(v + (\tilde{m} - \log K)/v) - KP(t, T)\Phi((\tilde{m} - \log K)/v)$$
$$= P(t, S)\Phi(v + (\tilde{m} - \log K)/v) - KP(t, T)\Phi((\tilde{m} - \log K)/v)$$
$$= P(t, S)\Phi\left(\frac{1}{v}\log\frac{P(t, S)}{KP(t, T)} + \frac{v}{2}\right) - KP(t, T)\Phi\left(\frac{1}{v}\log\frac{P(t, S)}{KP(t, T)} - \frac{v}{2}\right),$$

as in [Brigo and Mercurio (2006)], page 76. Note that this bond option price could also be computed from the joint law of $\left(r_T, \int_t^T r_s ds\right)$, which is Gaussian in the Vasicek case.

7.4 Inverse Change of Measure

For reference, in this section we give some computations of conditional inverse densities that will be useful later in Section 11.3 on swaption pricing.

Proposition 7.4. *We have*

$$\mathbb{E}_{\tilde{\mathbb{P}}}\left[\frac{d\mathbb{P}}{d\tilde{\mathbb{P}}}\Big| \mathcal{F}_t\right] = \frac{P(0, T)}{P(t, T)}\exp\left(\int_0^t r_s ds\right) \qquad 0 \leq t \leq T, \qquad (7.16)$$

and the process

$$t \longmapsto \frac{1}{P(t, T)}\exp\left(\int_0^t r_s ds\right), \qquad 0 \leq t \leq T,$$

is an $\mathcal{F}_t$-martingale under $\tilde{\mathbb{P}}$.

Proof. For all bounded and $\mathcal{F}_t$-measurable random variables F we have, using (7.4) and the characterization (12.4) of conditional expectation in Appendix A,

$$\mathbb{E}_{\tilde{\mathbb{P}}}\left[F\frac{d\mathbb{P}}{d\tilde{\mathbb{P}}}\right] = \mathbb{E}_{\mathbb{P}}[F]$$

$$= \mathbb{E}_{\mathbb{P}}\left[F\frac{P(t,T)}{P(t,T)}\right]$$

$$= \mathbb{E}_{\mathbb{P}}\left[\frac{F}{P(t,T)}\exp\left(-\int_t^T r_s ds\right)\right]$$

$$= \mathbb{E}_{\tilde{\mathbb{P}}}\left[F\frac{P(0,T)}{P(t,T)}\exp\left(\int_0^t r_s ds\right)\right],$$

where on the last line we applied Definition 7.1. This shows (7.16). $\square$

By Itô's calculus we also have from (7.7):

$$d\left(\frac{1}{P(t,T)}\right) = -\frac{1}{P(t,T)}r_t dt - \frac{1}{P(t,T)}\zeta_t(dB_t - \zeta_t dt),$$

and

$$d\left(\frac{1}{P(t,T)}\exp\left(\int_0^t r_s ds\right)\right) = -\frac{\zeta_t}{P(t,T)}\exp\left(\int_0^t r_s ds\right)(dB_t - \zeta_t dt)$$

$$= -\frac{1}{P(t,T)}\exp\left(\int_0^t r_s ds\right)\zeta_t d\tilde{B}_t,$$

which, from Proposition 7.3, recovers the second part of Proposition 7.4, i.e. the martingale property of

$$t \longmapsto \frac{1}{P(t,T)}\exp\left(\int_0^t r_s ds\right)$$

under $\tilde{\mathbb{P}}$.

7.5 Exercises

Exercise 7.1. (Exercise 6.1 continued).

(1) Derive a stochastic differential equation satisfied by $t \longmapsto P(t,T)$.
(2) Derive a stochastic differential equation satisfied by $t \longmapsto e^{-\int_0^t r_s ds}P(t,T)$.
(3) Express the conditional expectation

$$\mathbb{E}_{\mathbb{P}}\left[\frac{d\tilde{\mathbb{P}}}{d\mathbb{P}}\Big|\mathcal{F}_t\right]$$

in terms of $P(t,T)$, $P(0,T)$ and $e^{-\int_0^t r_s ds}$, $0 \le t \le T$.

(4) Find a stochastic differential equation satisfied by

$$t \longmapsto \mathbb{E}_\mathbb{P} \left[\frac{d\tilde{\mathbb{P}}}{d\mathbb{P}} \middle| \mathcal{F}_t \right].$$

(5) Compute the density $d\tilde{\mathbb{P}}/d\mathbb{P}$ of the forward measure with respect to $\mathbb{P}$ by solving the stochastic differential equation of question 4.
(6) Using the Girsanov Theorem 2.1, compute the dynamics of r_t under the forward measure.
(7) Compute the price

$$\mathbb{E}_\mathbb{P} \left[e^{-\int_t^T r_s ds} (P(T,S) - K)^+ \middle| \mathcal{F}_t \right] = P(t,T) \mathbb{E}_{\tilde{\mathbb{P}}} \left[(P(T,S) - K)^+ \middle| \mathcal{F}_t \right]$$

of a bond call option at time $t \geq 0$.

Exercise 7.2. (Exercise 6.2 continued).

(1) Compute the density

$$\frac{d\tilde{\mathbb{P}}}{d\mathbb{P}} = \frac{1}{P(0,T)} e^{-\int_0^T r_t dt}$$

of the forward measure $\tilde{\mathbb{P}}$ with respect to $\mathbb{P}$.
(2) Using the Girsanov Theorem 2.1, compute the dynamics of r_t under the forward measure.
(3) Assuming for simplicity that $b = 0$, compute the price

$$\mathbb{E}_\mathbb{P} \left[e^{-\int_0^T r_s ds} (P(T,S) - K)^+ \right] = P(0,T) \mathbb{E}_{\tilde{\mathbb{P}}} \left[(P(T,S) - K)^+ \right]$$

of a bond call option at time $t = 0$.

Exercise 7.3. We work in the Vasicek short rate model

$$dr_t = -br_t dt + \sigma dB_t,$$

where $(B_t)_{t \in \mathbb{R}_+}$ is a standard Brownian motion under $\mathbb{P}$.

(1) Recall the dynamics

$$\frac{dP(t,T_i)}{P(t,T_i)} = r_t dt + \zeta_t^i dB_t, \qquad i = 1, 2,$$

of the bond prices $P(t,T_1)$, $P(t,T_2)$ with maturities T_1, T_2.
(2) Recall the expression of the forward rate $f(t,T_1,T_2)$.

(3) Compute the dynamics of $f(t, T_1, T_2)$ under the forward measure $\mathbb{P}_2$ with

$$\frac{d\mathbb{P}_2}{d\mathbb{P}} = \frac{1}{P(0, T_2)} e^{-\int_0^{T_2} r_s ds}.$$

(4) Compute the price

$$(T_2 - T_1) \, \mathbb{E}\left[e^{-\int_t^{T_2} r_s ds}(f(T_1, T_1, T_2) - \kappa)^+ \Big| \mathcal{F}_t\right]$$

at time $t \in [0, T_1]$ of a caplet with strike κ using the expectation under the forward measure $\mathbb{P}_2$.

Exercise 7.4. Bond option hedging. (Exercise 4.3 continued). Consider a portfolio $(\xi_t^T, \xi_t^S)_{t \in [0,T]}$ made of two bonds with maturities T, S, and value

$$V_t = \xi_t^T P(t, T) + \xi_t^S P(t, S), \qquad 0 \le t \le T,$$

at time t, and assume that it *hedges* the bond call option payoff $(P(T, S) - \kappa)^+$, so that we have

$$V_t = \mathbb{E}\left[e^{-\int_t^T r_s ds}(P(T, S) - \kappa)^+ \Big| \mathcal{F}_t\right]$$

$$= P(t, T)\mathbb{E}_{\tilde{\mathbb{P}}}\left[(P(T, S) - \kappa)^+ \Big| \mathcal{F}_t\right], \qquad 0 \le t \le T.$$

(1) Assuming that $(\sigma_t^T)_{t \in [0,T]}$ and $(\sigma_t^S)_{t \in [0,S]}$ are deterministic functions, show that the price of a bond option with strike κ can be written as

$$\mathbb{E}\left[e^{-\int_t^T r_s ds}(P(T, S) - \kappa)^+ \Big| \mathcal{F}_t\right] = P(t, T)\, \mathbb{E}_{\tilde{\mathbb{P}}}\left[(P(T, S) - \kappa)^+ \Big| \mathcal{F}_t\right]$$

$$= P(t, T)C(X_t, \kappa, v(t, T)),$$

where X_t is the forward price $X_t := P(t, S)/P(t, T)$,

$$v^2(t, T) = \frac{1}{T - t} \int_t^T |\sigma_s^S - \sigma_s^T|^2 ds,$$

and $C(x, \kappa, \sigma)$ is a function to be determined. Recall that if X is a centered Gaussian random variable with mean m_t and variance v_t^2 given $\mathcal{F}_t$, we have

$$\mathbb{E}[(e^X - K)^+ \mid \mathcal{F}_t] = e^{m_t + v_t^2/2}\Phi\left(\frac{v_t}{2} + \frac{1}{v_t}(m_t + v_t^2/2 - \log K)\right) \quad (7.17)$$

$$-K\Phi\left(-\frac{v_t}{2} + \frac{1}{v_t}(m_t + v_t^2/2 - \log K)\right)$$

where $\Phi(x)$, $x \in \mathbb{R}$, denotes the Gaussian distribution function, cf. Lemma 2.3.

(2) We assume that the portfolio $(\xi_t^T, \xi_t^S)_{t\in[0,T]}$ is self-financing, i.e.

$$dV_t = \xi_t^T dP(t,T) + \xi_t^S dP(t,S), \qquad 0 \le t \le T. \qquad (7.18)$$

Show that the forward portfolio price $\hat{V}_t := V_t/P(t,T)$ satisfies

$$d\hat{V}_t = \frac{\partial C}{\partial x}(X_t, \kappa, v(t,T))dX_t. \qquad (7.19)$$

(3) Show that we have

$$dV_t = \left(\hat{V}_t - \frac{P(t,S)}{P(t,T)} \frac{\partial C}{\partial x}(X_t, \kappa, v(t,T)) \right) dP(t,T)$$

$$+ \frac{\partial C}{\partial x}(X_t, \kappa, v(t,T))dP(t,S).$$

(4) Compute the hedging portfolio strategy $(\xi_t^T, \xi_t^S)_{t\in[0,T]}$ of the bond call option on $P(T,S)$.

Exercise 7.5. (Exercise 5.3 continued).

(1) Compute the conditional density

$$\mathbb{E}\left[\frac{d\mathbb{P}_T}{d\mathbb{P}} \Big| \mathcal{F}_t \right] = \frac{P(t,T)}{P(0,T)} e^{-\int_0^t r_s^T ds}$$

of the forward measure $\mathbb{P}_T$ with respect to $\mathbb{P}$.

(2) Show that the process

$$\tilde{B}_t := B_t - \sigma t, \qquad 0 \le t \le T,$$

is a standard Brownian motion under $\mathbb{P}_T$.

(3) Compute the dynamics of X_t^S and $P(t,S)$ under $\mathbb{P}_T$.

Hint: Show that

$$-\mu(S-T) + \sigma(S-T) \int_0^t \frac{1}{S-s} dB_s = \frac{S-T}{S-t} \log P(t,S).$$

(4) Compute the bond option price

$$\mathbb{E}\left[e^{-\int_t^T r_s^T ds}(P(T,S)-K)^+ \Big| \mathcal{F}_t \right] = P(t,T) \mathbb{E}_T\left[(P(T,S)-K)^+ \Big| \mathcal{F}_t \right],$$

$$0 \le t < T < S.$$

Chapter 8

Curve Fitting and a Two-Factor Model

The short rate models considered in the previous chapters are one-factor models in the sense that their stochastic evolution is driven by a single Brownian motion. Such models pose several restriction to the fitting of forward curves, and they induce undesirable correlations in the price of zero-coupon bonds with different maturities. Our goal in this chapter is to study two-factor models which use two sources of randomness and allow for a wider choice of parameters for the fitting of forward curves.

8.1 Curve Fitting

Recall that in the Vasicek model, the instantaneous forward rate process is given by

$$f(t,T) = \frac{a}{b} - \frac{\sigma^2}{2b^2} + \left(r_t - \frac{a}{b} + \frac{\sigma^2}{b^2} \right) e^{-bx} - \frac{\sigma^2}{2b^2} e^{-2bx}, \qquad (8.1)$$

in the Musiela notation $(x = T - t)$. Hence

$$\frac{\partial f}{\partial T}(t,T) = e^{-b(T-t)} \left(-br_t + a - \frac{\sigma^2}{b} + \frac{\sigma^2}{b} e^{-b(T-t)} \right),$$

and one checks easily that the sign of the derivatives of f can only change once at most. As a consequence, the possible forward curves in the Vasicek model are limited to one change of "regime" per curve, as illustrated in Figure 8.1 for various values of r_t. For example, the short rate Vasicek paths of Figure 6.3 drive the instantaneous forward rates presented in Figure 8.2 which all converge to the "long rate"

$$\lim_{x \to \infty} f(t, t + x) = \lim_{T \to \infty} f(t,T) = \frac{a}{b} - \frac{\sigma^2}{2b^2}$$

as x goes to infinity.

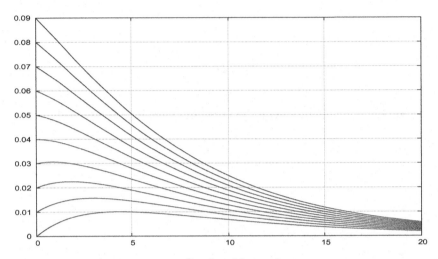

Fig. 8.1   Graphs of forward rates.

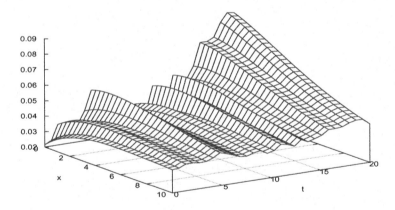

Fig. 8.2   Forward instantaneous curve $(t, x) \longmapsto f(t, t + x)$ in the Vasicek model.

From Relation (8.1) we know that the instantaneous forward curves in the Vasicek model "live" in the space of functions generated by

$$x \longmapsto z_1 + z_2 e^{z_3 x} + z_4 e^{z_5 x} \tag{8.2}$$

with

$$
\begin{cases}
z_1 = \dfrac{a}{b} - \dfrac{\sigma^2}{2b^2}, \\[2mm]
z_2 = r_t - \dfrac{a}{b} + \dfrac{\sigma^2}{b^2}, \\[2mm]
z_3 = -b, \\[2mm]
z_4 = -\dfrac{\sigma^2}{2b}, \\[2mm]
z_5 = -2b.
\end{cases}
$$

Unfortunately, this function space is contained neither in the Nelson-Siegel space, nor in the Svensson space. As previously noted in Chapter 6, typical realizations of such curves do not appear realistic enough to model the actual forward curves of e.g. Figure 6.4, and for this purpose, Svensson curves of the form

$$
x \longmapsto g(x) = z_1 + (z_2 + z_3 x)e^{-x z_4} + z_5 x e^{-x z_6}, \qquad x \geq 0,
$$

appear to yield a better modelization, cf. Figure 8.3 for a fit of the market data of Figure 6.4 using a Svensson curve.

Fig. 8.3 Comparison of market data *vs* a Svensson curve.

In order to solve this modelization problem one may think of constructing an instantaneous rate process taking values in the Svensson space, for example:

$$
\begin{aligned}
x \longmapsto f(t, T) &= f(t, t + x) \\
&= z_1(t) + (z_2(t) + z_3(t)x)e^{-x z_4(t)} + z_5(t)x e^{-x z_6(t)},
\end{aligned}
$$

$x \geq 0$, where $z_i(t)$, $i = 1, \ldots, 6$, are suitably chosen stochastic processes and $x = T - t$. An example of such a modelization has been given in Figure 6.1.

In this case the short rate may be defined as

$$r_t = f(t, t+0) = z_1(t) + z_2(t), \qquad t \in \mathbb{R}_+.$$

However this type of modelization raises the question whether it is consistent with the absence of arbitrage argument that led to the relation

$$P(t, T) = \exp\left(-\int_t^T f(t, s)ds\right) \tag{8.3}$$

and to the bond pricing relation

$$P(t, T) = \mathbb{E}\left[\exp\left(-\int_t^T r_s ds\right) \Big| \mathcal{F}_t\right], \tag{8.4}$$

which imply the condtion

$$\exp\left(-\int_t^T f(t, s)ds\right) = \mathbb{E}\left[\exp\left(-\int_t^T f(s, s)ds\right) \Big| \mathcal{F}_t\right].$$

The answer to that question is a priori negative since it is proved that the HJM curves cannot live in the Nelson-Siegel or Svensson spaces, cf. §3.5 of [Björk (2004)].

8.2   Deterministic Shifts

A possible way around the curve fitting problem is to use deterministic shifts for the fitting of one forward curve, usually at $t = 0$. Here we consider again a Vasicek model of forward rates in which we let $a = 0$ for simplicity. Let

$$f(t, T) := \varphi(T) + X_t e^{-b(T-t)} - \frac{\sigma^2}{2b^2}(1 - e^{-b(T-t)})^2, \tag{8.5}$$

where $T \longmapsto \varphi(T)$ is given deterministic function, and X_t is the solution of

$$dX_t = -bX_t dt + \sigma dB_t$$

with $X_0 = 0$.

From Relation (8.5) at $t = 0$ we check that choosing $T \longmapsto \varphi(T)$ as

$$\varphi(T) := f^M(0, T) + \frac{\sigma^2}{2b^2}(1 - e^{-bT})^2 \tag{8.6}$$

allows one to match any initial market term structure $T \longmapsto f^M(0,T)$, or any (fixed) term structure of our choice at a given (unique) time t, by the equality $f(0,T) = f^M(0,T)$, $T \geq 0$.

The interest in this method is to achieve consistency with the absence of arbitrage condition (4.2) and Relations (8.3) and (8.4) above by defining the short rate r_t as

$$r_t = f(t,t) = \varphi(t) + X_t. \tag{8.7}$$

Indeed, in this case we have

$$P(t,T) = \mathbb{E}\left[\exp\left(-\int_t^T r_s ds\right) \Big| \mathcal{F}_t\right]$$

$$= \mathbb{E}\left[\exp\left(-\int_t^T \varphi(s)ds - \int_t^T X_s ds\right) \Big| \mathcal{F}_t\right]$$

$$= \exp\left(-\int_t^T \varphi(s)ds\right) \mathbb{E}\left[\exp\left(-\int_t^T X_s ds\right) \Big| \mathcal{F}_t\right]$$

$$= \exp\left(-\int_t^T \varphi(s)ds\right) \exp\left(-\int_t^T \left(X_t e^{-b(s-t)} - \frac{\sigma^2}{2b^2}(1 - e^{-b(s-t)})^2\right) ds\right)$$

$$= \exp\left(-\int_t^T f(t,s)ds\right),$$

where we used the expression

$$\mathbb{E}\left[\exp\left(-\int_t^T X_s ds\right) \Big| \mathcal{F}_t\right]$$

$$= \exp\left(-\int_t^T \left(X_t e^{-b(s-t)} - \frac{\sigma^2}{2b^2}(1 - e^{-b(s-t)})^2\right) ds\right)$$

of the bond price as the exponential of an integral of the instantaneous forward rate in a Vasicek model with $a = 0$. Recall that this argument is however restricted to the fitting of a single (initial) curve.

8.3 The Correlation Problem

The correlation problem is another issue of concern when using the affine models considered so far. Let us consider the following bond price simulation with maturity $T_3 = 30$, cf. Figure 8.4. Next, the simulation of another

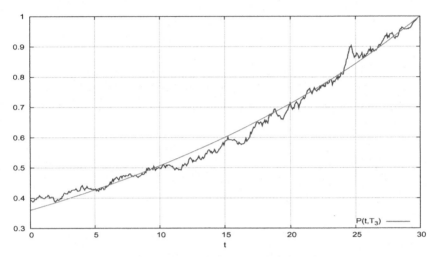

Fig. 8.4   Graph of $t \longmapsto P(t, T_3)$.

Fig. 8.5   Graph of $t \longmapsto P(t, T_2)$.

bond with maturity $T_2 = 20$ is given in Figure 8.5. Let us compare these graphs together with a simulation of a bond with maturity $T_1 = 10$ as given in Figure 8.6. Clearly, the bond prices $P(t, T_1)$ and $P(t, T_2)$ with maturities T_1 and T_2 are linked by the relation

$$P(t, T_2) = P(t, T_1) \exp(A(t, T_2) - A(t, T_1) + r_t(C(t, T_2) - C(t, T_1))),$$

meaning that bond prices with different maturities could be deduced from each other, which is unrealistic.

Note that if X and Y are two linear combinations of a same random variable Z:

$$X = a + bZ, \qquad Y = c + dZ,$$

then their covariance (see Appendix A) equals

$$\begin{aligned} \mathrm{Cov}(X, Y) &= \mathrm{Cov}(a + bZ, c + dZ) \\ &= \mathrm{Cov}(bZ, dZ) \\ &= bd \, \mathrm{Cov}(Z, Z) \\ &= bd \, \mathrm{Var}(Z), \end{aligned}$$

and

$$\mathrm{Var}\, X = b^2 \, \mathrm{Var}\, Z, \qquad \mathrm{Var}\, Y = d^2 \, \mathrm{Var}\, Z,$$

hence

$$\mathrm{Cor}(X, Y) = \frac{\mathrm{Cov}(X, Y)}{\sqrt{\mathrm{Var}\, X}\sqrt{\mathrm{Var}\, Y}} = 1,$$

i.e. X and Y are completely correlated. This property holds in particular for the affine models of short rates, meaning that in affine models we have the perfect correlation

$$\mathrm{Cor}(\log P(t, T_1), \log P(t, T_2)) = 1.$$

Fig. 8.6 Graph of $t \longmapsto P(t, T_1)$.

8.4   Two-Factor Model

A partial solution to the correlation problem would be to consider two control processes $(X_t)_{t\in\mathbb{R}_+}$, $(Y_t)_{t\in\mathbb{R}_+}$ which are solution of

$$\begin{cases} dX_t = \mu_1(t, X_t)dt + \sigma_1(t, X_t)dB_t^{(1)}, \\ \\ dY_t = \mu_2(t, Y_t)dt + \sigma_2(t, Y_t)dB_t^{(2)}, \end{cases} \tag{8.8}$$

where $(B_t^{(1)})_{t\in\mathbb{R}_+}$, $(B_t^{(2)})_{t\in\mathbb{R}_+}$ are two (possibly dependent) Brownian motions, and to define the associated bond prices as

$$P(t, T_1) = \mathbb{E}\left[\exp\left(-\int_t^{T_1} X_s ds\right) \Big| \mathcal{F}_t\right]$$

and

$$P(t, T_2) = \mathbb{E}\left[\exp\left(-\int_t^{T_2} Y_s ds\right) \Big| \mathcal{F}_t\right].$$

This type of approach would nevertheless lead to other difficulties, namely:

- the presence of two candidate processes X_t and Y_t to represent the interest rate,

- the need to introduce a new control process for each additional maturity T_n, $n \geq 3$.

In the sequel we will assume that $(B_t^{(1)})_{t\in\mathbb{R}_+}$ and $(B_t^{(2)})_{t\in\mathbb{R}_+}$ have correlation $\rho \in [-1, 1]$, that is

$$\text{Cov}(B_s^{(1)}, B_t^{(2)}) = \rho \min(s, t), \qquad s, t \in \mathbb{R}_+. \tag{8.9}$$

Note that in terms of stochastic differentials this implies

$$dB_t^{(1)} dB_t^{(2)} = \rho dt. \tag{8.10}$$

In practice, $(B^{(1)})_{t\in\mathbb{R}_+}$ and $(B^{(2)})_{t\in\mathbb{R}_+}$ can be constructed from two independent Brownian motions $(W^{(1)})_{t\in\mathbb{R}_+}$ and $(W^{(2)})_{t\in\mathbb{R}_+}$, by letting

$$\begin{cases} B_t^{(1)} = W_t^{(1)}, \\ \\ B_t^{(2)} = \rho W_t^{(1)} + \sqrt{1 - \rho^2} W_t^{(2)}, \qquad t \in \mathbb{R}_+, \end{cases}$$

and Relations (8.9) and (8.10) are easily satisfied from this construction.

In two-factor models one chooses to build the short term interest rate r_t via

$$r_t = \varphi(t) + X_t + Y_t, \qquad t \in \mathbb{R}_+,$$

where the function $\varphi(T)$ can be chosen to fit the initial forward curve as in (8.6).

Following the standard arbitrage arguments of Chapter 4, we define the price of a bond with maturity T as

$$P(t,T) := \mathbb{E}\left[\exp\left(-\int_t^T r_s ds\right)\Big|\mathcal{F}_t\right]. \tag{8.11}$$

Being a solution to the stochastic differential equation (8.8), the couple $(X_t, Y_t)_{t\in\mathbb{R}_+}$ is Markovian and we can write

$$P(t,T) = \mathbb{E}\left[\exp\left(-\int_t^T r_s ds\right)\Big|X_t,\ Y_t\right], \tag{8.12}$$

which however does no longer directly depend on the short rate r_t.

Nevertheless, $P(t,T)$ can be written as a function

$$P(t,T) = F(t, X_t, Y_t)$$

of t, X_t and Y_t, and one may use the Itô formula with two variables to derive a PDE on $\mathbb{R}^2$ for the bond price $P(t,T)$, using the fact that

$$t \longmapsto e^{-\int_0^t r_s ds} P(t,T) = e^{-\int_0^t r_s ds}\, \mathbb{E}\left[\exp\left(-\int_t^T r_s ds\right)\Big|\mathcal{F}_t\right]$$

$$= \mathbb{E}\left[\exp\left(-\int_0^T r_s ds\right)\Big|\mathcal{F}_t\right]$$

is an $\mathcal{F}_t$-martingale under $\mathbb{P}$.

We have

$$d\left(e^{-\int_0^t r_s ds} P(t,T)\right) = -r_t e^{-\int_0^t r_s ds} P(t,T)dt + e^{-\int_0^t r_s ds} dP(t,T) \tag{8.13}$$

$$= -r_t e^{-\int_0^t r_s ds} P(t,T)dt + e^{-\int_0^t r_s ds} dF(t, X_t, Y_t)$$

$$= -r_t e^{-\int_0^t r_s ds} P(t,T)dt + e^{-\int_0^t r_s ds}\frac{\partial F}{\partial x}(t, X_t, Y_t)dX_t + \frac{\partial F}{\partial t}(t, X_t, Y_t)dt$$

$$+ e^{-\int_0^t r_s ds}\frac{\partial F}{\partial y}(t, X_t, Y_t)dY_t + \frac{1}{2}e^{-\int_0^t r_s ds}\frac{\partial^2 F}{\partial x^2}(t, X_t, Y_t)\sigma_1^2(t, X_t)dt$$

$$+\frac{1}{2}e^{-\int_0^t r_s ds}\frac{\partial^2 F}{\partial y^2}(t, X_t, Y_t)\sigma_2^2(t, Y_t)dt$$

$$+e^{-\int_0^t r_s ds}\rho\frac{\partial^2 F}{\partial x \partial y}(t, X_t, Y_t)\sigma_1(t, X_t)\sigma_2(t, Y_t)dt$$

$$= e^{-\int_0^t r_s ds}\frac{\partial F}{\partial x}(t, X_t, Y_t)\sigma_1(t, X_t)dB_t^{(1)}$$

$$+e^{-\int_0^t r_s ds}\frac{\partial F}{\partial y}(t, X_t, Y_t)\sigma_2(t, Y_t)dB_t^{(2)}$$

$$e^{-\int_0^t r_s ds}\left(-r_t P(t, T) + \frac{\partial F}{\partial x}(t, X_t, Y_t)\mu_1(t, X_t) + \frac{\partial F}{\partial y}(t, X_t, Y_t)\mu_2(t, Y_t)\right.$$

$$+\frac{1}{2}\frac{\partial^2 F}{\partial x^2}(t, X_t, Y_t)\sigma_1^2(t, X_t) + \frac{1}{2}\frac{\partial^2 F}{\partial y^2}(t, X_t, Y_t)\sigma_2^2(t, Y_t)$$

$$\left.+\rho\frac{\partial^2 F}{\partial x \partial y}(t, X_t, Y_t)\sigma_1(t, X_t)\sigma_2(t, Y_t) + \frac{\partial F}{\partial t}(t, X_t, Y_t)dt\right) dt,$$

hence the bond pricing PDE is

$$-(\varphi(t) + x + y)F(t, x, y) + \mu_1(t, x)\frac{\partial F}{\partial x}(t, x, y) + \mu_2(t, y)\frac{\partial F}{\partial y}(t, x, y)$$

$$+\frac{1}{2}\sigma_1^2(t, x)\frac{\partial^2 F}{\partial x^2}(t, x, y) + \frac{1}{2}\sigma_2^2(t, y)\frac{\partial^2 F}{\partial y^2}(t, x, y)$$

$$+\rho\sigma_1(t, x)\sigma_2(t, y)\frac{\partial^2 F}{\partial x \partial y}(t, x, y) + \frac{\partial F}{\partial t}(t, X_t, Y_t) = 0. \qquad (8.14)$$

Next we consider another Vasicek-type example where

$$\begin{cases} dX_t = -aX_t dt + \sigma dB_t^{(1)}, \\ \\ dY_t = -bY_t dt + \eta dB_t^{(2)}. \end{cases} \qquad (8.15)$$

Here, instead of attempting to solve the 2-dimensional PDE (8.14) we choose to compute $P(t, T) = F(t, X_t, Y_t)$ from its expression (8.12) as a conditional expectation.

Proposition 8.1. *We have*

$$P(t, T) = \exp\left(-\int_t^T \varphi(s)ds - \frac{1}{a}(1 - e^{-a(T-t)})X_t - \frac{1}{b}(1 - e^{-b(T-t)})Y_t\right)$$

$$\times \exp\left(\frac{\sigma^2}{2a^2}\int_t^T (e^{-a(T-s)} - 1)^2 ds + \frac{\eta^2}{2b^2}\int_t^T (e^{-b(T-s)} - 1)^2 ds\right)$$

$$\times \exp\left(\rho \frac{\sigma\eta}{ab}\int_t^T (e^{-a(T-s)}-1)(e^{-b(T-s)}-1)ds\right),$$

$0 \le t \le T$.

Proof. See [Brigo and Mercurio (2006)], Chapter 4, Appendix A, for a different proof. Note that we have

$$\int_0^t X_s ds = \frac{1}{a}\left(\sigma B_t^{(1)} - X_t\right)$$

$$= \frac{\sigma}{a}\left(B_t^{(1)} - \int_0^t e^{-a(t-s)}dB_s^{(1)}\right)$$

$$= \frac{\sigma}{a}\int_0^t (1-e^{-a(t-s)})dB_s^{(1)},$$

hence

$$\int_t^T X_s ds = \int_0^T X_s ds - \int_0^t X_s ds$$

$$= \frac{\sigma}{a}\int_0^T (1-e^{-a(T-s)})dB_s^{(1)} - \frac{\sigma}{a}\int_0^t (1-e^{-a(t-s)})dB_s^{(1)}$$

$$= -\frac{\sigma}{a}\left(\int_0^t (e^{-a(T-s)}-e^{-a(t-s)})dB_s^{(1)} + \int_t^T (e^{-a(T-s)}-1)dB_s^{(1)}\right)$$

$$= -\frac{\sigma}{a}(e^{-a(T-t)}-1)\int_0^t e^{-a(t-s)}dB_s^{(1)} - \frac{\sigma}{a}\int_t^T (e^{-a(T-s)}-1)dB_s^{(1)}$$

$$= -\frac{1}{a}(e^{-a(T-t)}-1)X_t - \frac{\sigma}{a}\int_t^T (e^{-a(T-s)}-1)dB_s^{(1)},$$

and similarly,

$$\int_t^T Y_s ds = -\frac{1}{b}(e^{-b(T-t)}-1)Y_t - \frac{\eta}{b}\int_t^T (e^{-b(T-s)}-1)dB_s^{(2)}.$$

Hence, conditionally to $\mathcal{F}_t$, the random vector $\left(\int_t^T X_s ds, \int_t^T Y_s ds\right)$ is Gaussian with mean

$$\begin{pmatrix} \mathbb{E}\left[\int_t^T X_s ds \middle| \mathcal{F}_t\right] \\ \\ \mathbb{E}\left[\int_t^T Y_s ds \middle| \mathcal{F}_t\right] \end{pmatrix} = \begin{pmatrix} \frac{1}{a}(1-e^{-a(T-t)})X_t \\ \\ \frac{1}{b}(1-e^{-b(T-t)})Y_t \end{pmatrix}$$

and conditional covariance matrix

$$
\mathrm{Cov}\left[\int_t^T X_s ds, \int_t^T Y_s ds \Big| \mathcal{F}_t\right] =
$$

$$
\begin{pmatrix}
\dfrac{\sigma^2}{a^2}\displaystyle\int_t^T (e^{-a(T-s)}-1)^2 ds & \rho\dfrac{\sigma\eta}{ab}\displaystyle\int_t^T (e^{-a(T-s)}-1)(e^{-b(T-s)}-1)ds \\[4mm]
\rho\dfrac{\sigma\eta}{ab}\displaystyle\int_t^T (e^{-a(T-s)}-1)(e^{-b(T-s)}-1)ds & \dfrac{\eta^2}{b^2}\displaystyle\int_t^T (e^{-b(T-s)}-1)^2 ds
\end{pmatrix}
$$

obtained by the Itô isometry (1.4). The variance of $\int_t^T X_s ds + \int_t^T Y_s ds$ given $\mathcal{F}_t$ equals

$$
\left\langle \mathrm{Cov}\left[\int_t^T X_s ds, \int_t^T Y_s ds \Big| \mathcal{F}_t\right]\begin{bmatrix}1\\1\end{bmatrix}, \begin{bmatrix}1\\1\end{bmatrix}\right\rangle_{\mathbb{R}^2}
$$

$$
= \mathrm{Var}\left[\int_t^T X_s ds \Big| \mathcal{F}_t\right] + \mathrm{Var}\left[\int_t^T Y_s ds \Big| \mathcal{F}_t\right] + 2\,\mathrm{Cov}\left(\int_t^T X_s ds, \int_t^T Y_s ds \Big| \mathcal{F}_t\right),
$$

and we have

$$
P(t,T) = \mathbb{E}\left[\exp\left(-\int_t^T r_s ds\right)\Big| \mathcal{F}_t\right]
$$

$$
= \exp\left(-\int_t^T \varphi(s)ds\right)\mathbb{E}\left[\exp\left(-\int_t^T X_s ds - \int_t^T Y_s ds\right)\Big| \mathcal{F}_t\right]
$$

$$
= \exp\left(-\int_t^T \varphi(s)ds - \mathbb{E}\left[\int_t^T X_s ds \Big| \mathcal{F}_t\right] - \mathbb{E}\left[\int_t^T Y_s ds \Big| \mathcal{F}_t\right]\right)
$$

$$
\times \exp\left(\frac{1}{2}\left\langle \mathrm{Cov}\left[\int_t^T X_s ds, \int_t^T Y_s ds \Big| \mathcal{F}_t\right]\begin{bmatrix}1\\1\end{bmatrix}, \begin{bmatrix}1\\1\end{bmatrix}\right\rangle_{\mathbb{R}^2}\right)
$$

$$
= \exp\left(-\int_t^T \varphi(s)ds - \frac{1}{a}(1-e^{-a(T-t)})X_t - \frac{1}{b}(1-e^{-b(T-t)})Y_t\right)
$$

$$
\times \exp\left(\frac{\sigma^2}{2a^2}\int_t^T (e^{-a(T-s)}-1)^2 ds + \frac{\eta^2}{2b^2}\int_t^T (e^{-b(T-s)}-1)^2 ds\right)
$$

$$
\times \exp\left(\rho\frac{\sigma\eta}{ab}\int_t^T (e^{-a(T-s)}-1)(e^{-b(T-s)}-1)ds\right),
$$

see Appendix A for details on the Laplace transform (12.3) of Gaussian random vectors. □

The above proposition shows in particular that the solution to the 2-dimensional PDE (8.14) is

$$F(t, x, y) = \exp\left(-\int_t^T \varphi(s)ds - \frac{1}{a}(1 - e^{-a(T-t)})x - \frac{1}{b}(1 - e^{-b(T-t)})y\right)$$

$$\times \exp\left(\frac{\sigma^2}{2a^2}\int_t^T (e^{-a(T-s)} - 1)^2 ds + \frac{\eta^2}{2b^2}\int_t^T (e^{-b(T-s)} - 1)^2 ds\right)$$

$$\times \exp\left(\rho\frac{\sigma\eta}{ab}\int_t^T (e^{-a(T-s)} - 1)(e^{-b(T-s)} - 1)ds\right).$$

The bond price $P(t, T)$ can also be written as

$$P(t, T) = F_1(t, X_t)F_2(t, Y_t)\exp\left(-\int_t^T \varphi(s)ds + U(t, T)\right), \qquad (8.16)$$

where $F_1(t, X_t)$ and $F_2(t, Y_t)$ are the bond prices associated to X_t and Y_t in the Vasicek model:

$$F_1(t, X_t) = \mathbb{E}\left[\exp\left(-\int_t^T X_s ds\right)\Big| X_t\right]$$

$$= \exp\left(\frac{\sigma^2}{a^2}\left(T - t + \frac{2}{a}e^{-a(T-t)} - \frac{1}{2a}e^{-2a(T-t)} - \frac{3}{2a}\right) - \frac{1 - e^{-a(T-t)}}{a}X_t\right),$$

$$F_2(t, Y_t) = \mathbb{E}\left[\exp\left(-\int_t^T Y_s ds\right)\Big| Y_t\right]$$

$$= \exp\left(\frac{\eta^2}{b^2}\left(T - t + \frac{2}{b}e^{-b(T-t)} - \frac{1}{2b}e^{-2b(T-t)} - \frac{3}{2b}\right) - \frac{1 - e^{-b(T-t)}}{b}Y_t\right),$$

and

$$U(t, T) = \rho\frac{\sigma\eta}{ab}\left(T - t + \frac{e^{-a(T-t)} - 1}{a} + \frac{e^{-b(T-t)} - 1}{b} - \frac{e^{-(a+b)(T-t)} - 1}{a + b}\right),$$

$0 \le t \le T$, is a correlation term which vanishes when $(B_t^{(1)})_{t\in\mathbb{R}_+}$ and $(B_t^{(2)})_{t\in\mathbb{R}_+}$ are independent, i.e. when $\rho = 0$.

Partial differentiation of $\log P(t, T)$ with respect to T leads to the instantaneous forward rate

$$f(t, T) = -\frac{\partial \log P(t, T)}{\partial T}$$

$$= \varphi(T) + f_1(t, T) + f_2(t, T) - \rho\frac{\sigma\eta}{ab}(1 - e^{-a(T-t)})(1 - e^{-b(T-t)})$$

$$= \varphi(T) + X_t e^{-a(T-t)} - \frac{\sigma^2}{2a^2}(1 - e^{-a(T-t)})^2 + Y_t e^{-b(T-t)} \quad (8.17)$$

$$- \frac{\eta^2}{2b^2}(1 - e^{-b(T-t)})^2 - \rho\frac{\sigma\eta}{ab}(1 - e^{-a(T-t)})(1 - e^{-b(T-t)}),$$

where $f_1(t,T)$, $f_2(t,T)$ are the instantaneous forward rates corresponding to X_t and Y_t respectively, i.e.

$$f_1(t,T) = X_t e^{-a(T-t)} - \frac{\sigma^2}{2a^2}(1 - e^{-a(T-t)})^2$$

and

$$f_2(t,T) = Y_t e^{-b(T-t)} - \frac{\eta^2}{2b^2}(1 - e^{-b(T-t)})^2.$$

Clearly, the forward instantaneous rate now depends on a larger number of degrees of freedom, in particular one is now allowed to choose independently the coefficients a and b appearing in the exponentials in (8.17). An example of a forward rate curve obtained in this way is given in Figure 8.7.

Fig. 8.7   Graph of forward rates in a two-factor model.

Next in Figure 8.8 we present a graph of the evolution of forward curve in a two-factor model.

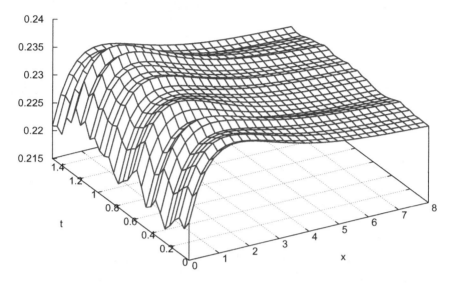

Fig. 8.8 Random evolution of forward rates in a two-factor model.

8.5 Exercises

Exercise 8.1. Find the stochastic differential equation satisfied by $P(t,T)$ defined in (8.16).

Problem 8.2. Consider the Hull-White model in which the short term interest rate process $(r_t)_{t \in \mathbb{R}_+}$ satisfies the equation

$$dr_t = (\theta(t) - ar_t)dt + \sigma dB_t, \qquad (8.18)$$

where $a \in \mathbb{R}$, $\theta(t)$ is a deterministic function of t, the initial condition r_0 is deterministic, and $(B_t)_{t \in \mathbb{R}_+}$ is a standard Brownian motion under $\mathbb{P}$, generating the filtration $(\mathcal{F}_t)_{t \in \mathbb{R}_+}$. Let the bond price $P(t,T)$ be defined under the absence of arbitrage hypothesis as

$$P(t,T) = \mathbb{E}\left[e^{-\int_t^T r_s ds} \Big| \mathcal{F}_t\right], \qquad 0 \le t \le T.$$

Recall that from the Markov property of $(r_t)_{t \in \mathbb{R}_+}$, there exists a function

$$(t,x) \longmapsto F(t,x)$$

such that

$$F(t,r_t) = P(t,T), \qquad 0 \le t \le T.$$

(1) Let $(X_t)_{t \in \mathbb{R}_+}$ denote the solution of the stochastic differential equation

$$\begin{cases} dX_t = -aX_t dt + \sigma dB_t, & t > 0, \\\\ X_0 = 0. \end{cases} \qquad (8.19)$$

Show that

$$r_t = r_0 e^{-at} + \varphi(t) + X_t, \qquad t > 0,$$

where

$$\varphi(t) = \int_0^t \theta(u) e^{-a(t-u)} du, \qquad t \in \mathbb{R}_+.$$

(2) Using Itô's calculus, derive the PDE satisfied by the function $(t, x) \longmapsto F(t, x)$.

(3) Recall (cf. Exercise 4.5) that $\int_t^T X_s ds$ has a Gaussian distribution given $\mathcal{F}_t$, with

$$\mathbb{E}\left[\int_t^T X_s ds \Big| \mathcal{F}_t \right] = \frac{X_t}{a}(1 - e^{-a(T-t)})$$

and

$$\operatorname{Var}\left[\int_t^T X_s ds \Big| \mathcal{F}_t \right] = \frac{\sigma^2}{a^2} \int_t^T (e^{-a(T-s)} - 1)^2 ds.$$

Show that the bond price $P(t, T)$ can be written as

$$P(t, T) = e^{A(t,T) + X_t C(t,T)},$$

where $A(t, T)$ and $C(t, T)$ are functions to be determined.

(4) Show that in this model, the instantaneous forward rate

$$f(t, T) = -\frac{\partial \log P(t, T)}{\partial T}$$

is given by

$$f(t, T) = r_0 e^{-aT} + \varphi(T) + X_t e^{-a(T-t)} - \frac{\sigma^2}{2a^2}(1 - e^{-a(T-t)})^2, \quad 0 \le t \le T.$$

(5) Compute $d_t f(t, T)$ and derive the stochastic equation satisfied by the instantaneous forward rate $f(t, T)$.

(6) Check that the HJM absence of arbitrage condition is satisfied in this equation.

(7) Let the market data of an initial interest rate curve be given by a function

$$T \longmapsto f^M(0,T).$$

Show that the function $\varphi(t)$ can be chosen in such a way that the theoretical value $f(0,T)$ matches the market data $f^M(0,T)$, i.e.

$$f(0,T) = f^M(0,T), \qquad \forall\, T > 0.$$

(8) Show that choosing $\theta(t)$ equal to

$$\theta(t) = af^M(0,t) + \frac{\partial f^M}{\partial t}(0,t) + \frac{\sigma^2}{2a}(1 - e^{-2at}), \qquad t > 0,$$

entails

$$f(0,T) = f^M(0,T), \qquad \forall\, T > 0.$$

(9) Show that we have

$$\frac{dP(t,T)}{P(t,T)} = r_t dt + \zeta_t dB_t,$$

and

$$d\left(e^{-\int_0^t r_s ds} P(t,T) \right) = \zeta_t e^{-\int_0^t r_s ds} P(t,T) dB_t, \qquad (8.20)$$

where $(\zeta_t)_{t \in [0,T]}$ is a process to be determined.

(10) Let the forward measure $\tilde{\mathbb{P}}$ be defined via its density

$$\mathbb{E}\left[\frac{d\tilde{\mathbb{P}}}{d\mathbb{P}} \Big| \mathcal{F}_t \right] = \frac{P(t,T)}{P(0,T)} e^{-\int_0^t r_s ds}$$

with respect to $\mathbb{P}$, $0 \le t \le T$. Compute $d\tilde{\mathbb{P}}/d\mathbb{P}$ by solving Equation (8.20).

(11) Using the Girsanov Theorem 2.1, compute the dynamics of r_t under the forward measure $\tilde{\mathbb{P}}$.

(12) Using Itô's calculus, show that

$$t \longmapsto \frac{P(t,S)}{P(t,T)}, \qquad 0 \le t \le T \le S,$$

is a martingale under $\tilde{\mathbb{P}}$.

(13) Show that

$$\mathbb{E}_{\tilde{\mathbb{P}}}\left[P(T,S) \big| \mathcal{F}_t \right] = \frac{P(t,S)}{P(t,T)}, \qquad 0 \le t \le T \le S,$$

and from this identity, deduce the value of

$$A(T,S) + C(T,S)\,\mathbb{E}[X_T \mid \mathcal{F}_t] + \frac{1}{2}|C(T,S)|^2 \operatorname{Var}[X_T \mid \mathcal{F}_t]$$

in terms of $P(t,S)/P(t,T)$.

(14) Compute the price

$$\mathbb{E}_{\mathbb{P}}\left[e^{-\int_t^T r_s ds}(K - P(T,S))^+ \Big| \mathcal{F}_t\right] = P(t,T)\,\mathbb{E}_{\tilde{\mathbb{P}}}\left[(K - P(T,S))^+ \Big| \mathcal{F}_t\right]$$

at time t of a bond *put* option.

Recall that $x - K = (x - K)^+ - (K - x)^+$, and that if X is a Gaussian random variable with mean m_t and variance v_t^2 given $\mathcal{F}_t$, we have

$$\mathbb{E}[(e^X - K)^+ \mid \mathcal{F}_t] = \Phi\left(\frac{v_t}{2} + \frac{1}{v_t}(m_t + v_t^2/2 - \log K)\right)$$
$$-K\Phi\left(-\frac{v_t}{2} + \frac{1}{v_t}(m_t + v_t^2/2 - \log K)\right)$$

where

$$\Phi(x) = \int_{-\infty}^x e^{-y^2/2}\frac{dy}{\sqrt{2\pi}}, \qquad x \in \mathbb{R},$$

denotes the Gaussian cumulative distribution function.

Chapter 9

A Credit Default Model

The bond models studied in the previous chapters rely on the hypothesis that the bond payoff at maturity is always equal to $1, thus excluding the possibility of default. In this chapter we study a bond model in which default can occur at a random time τ. We also consider the associated options (credit default swaps) that are designed as a protection against default.

9.1 Survival Probabilities

Let τ denote the random lifetime of an entity, and let $\mathbb{P}(\tau \geq t)$, $t > 0$, denote its probability of surviving at least t years. The conditional probability for that entity to survive up to time T, given that it has survived up to time t is given by

$$\mathbb{P}(\tau > T \mid \tau > t) = \frac{\mathbb{P}(\tau > T \text{ and } \tau > t)}{\mathbb{P}(\tau > t)}$$

$$= \frac{\mathbb{P}(\tau > T)}{\mathbb{P}(\tau > t)}, \qquad 0 \leq t \leq T.$$

Letting

$$\lambda(t) := \frac{\mathbb{P}(\tau < t + dt \mid \tau > t)}{dt}$$

$$= \frac{1}{\mathbb{P}(\tau > t)} \frac{\mathbb{P}(\tau < t + dt \text{ and } \tau > t)}{dt}$$

$$= \frac{1}{\mathbb{P}(\tau > t)} \frac{\mathbb{P}(\tau > t) - \mathbb{P}(\tau > t + dt)}{dt}$$

$$= -\frac{d}{dt} \log \mathbb{P}(\tau > t)$$

$$= -\frac{1}{\mathbb{P}(\tau > t)}\frac{d}{dt}\mathbb{P}(\tau > t), \qquad t > 0,$$

we get the differential equation

$$\frac{d}{dt}\mathbb{P}(\tau > t) = -\lambda(t)\mathbb{P}(\tau > t),$$

with initial condition $\mathbb{P}(\tau > 0) = 1$, and solution

$$\mathbb{P}(\tau > t) = \exp\left(-\int_0^t \lambda(u)du\right), \qquad t \in \mathbb{R}_+. \tag{9.1}$$

This implies that

$$\mathbb{P}(\tau > T \mid \tau > t) = \frac{\mathbb{P}(\tau > T)}{\mathbb{P}(\tau > t)} = \exp\left(-\int_t^T \lambda(u)du\right), \qquad 0 \le t \le T.$$

Given survival up to time t, survival over the short time period $[t, t + dt]$ occurs with probability

$$\mathbb{P}(\tau > t + dt \mid \tau > t) = e^{-\lambda(t)dt} \simeq 1 - \lambda(t)dt, \tag{9.2}$$

and default occurs on $[t, t + dt]$ with probability

$$\mathbb{P}(\tau < t + dt \mid \tau > t) = 1 - e^{-fdt} \simeq \lambda(t)dt, \tag{9.3}$$

as dt tends to 0. The above formulas can be recovered informally as

$$\mathbb{P}(\tau > T) = \prod_{0 < t < T} \mathbb{P}(\tau > t + dt \mid \tau > t)$$

$$= \prod_{0 < t < T} \exp\left(-\lambda(t)dt\right)$$

$$= \exp\left(-\int_0^T \lambda(t)dt\right).$$

When $\lambda(t) = \lambda > 0$ is a constant function, the random variable τ has the exponential distribution with parameter λ, i.e.

$$\mathbb{P}(\tau > T) = e^{-\lambda T}, \qquad T \ge 0.$$

Given $(\tau_n)_{n \ge 1}$ a sequence of i.i.d. exponentially distributed random variables, letting

$$T_n = \tau_1 + \cdots + \tau_n, \qquad n \ge 1,$$

and

$$N_t = \sum_{n \ge 1} \mathbf{1}_{[T_n, \infty)}(t)$$

defines a standard Poisson process $(N_t)_{t \in \mathbb{R}_+}$ with intensity $\lambda > 0$, with

$$\mathbb{P}(N_t - N_s = k) = e^{-\lambda(t-s)} \frac{\lambda^k (t-s)^k}{k!}, \qquad k \geq 0.$$

The intensity of the process can in fact made time-dependent (e.g. by a time change), i.e.

$$\mathbb{P}(N_t - N_s = k) = \exp\left(-\int_s^t \lambda(u)du\right) \frac{\left(\int_s^t \lambda(u)du\right)^k}{k!}, \qquad k \geq 0.$$

In particular,

$$\mathbb{P}(N_{t+dt} - N_t = k) = \begin{cases} e^{-\lambda(t)dt} \simeq 1 - \lambda(t)dt, & k = 0, \\[2mm] \lambda(t)e^{-\lambda(t)dt}dt \simeq \lambda(t)dt, & k = 1, \\[2mm] o(dt), & k \geq 2, \end{cases}$$

and $\mathbb{P}(N_{t+dt} - N_t = 0)$, $\mathbb{P}(N_{t+dt} - N_t = 1)$ coincide respectively with (9.2) and (9.3) above. The intensity process $(\lambda(t))_{t \in \mathbb{R}_+}$ can also be made random in the case of Cox processes.

9.2 Stochastic Default

We now assume that the default rate $(\lambda_t)_{t \in \mathbb{R}_+}$ is a random process which is adapted to a filtration $(\mathcal{F}_t)_{t \in \mathbb{R}_+}$.

The random time τ is called a *stopping time* with respect to $(\mathcal{F}_t)_{t \in \mathbb{R}_+}$ if we have

$$\{\tau > t\} \in \mathcal{F}_t, \qquad t \in \mathbb{R}_+,$$

i.e. the knowledge of whether default already occurred at time t is contained in $\mathcal{F}_t$, $t \in \mathbb{R}_+$. In this case we have

$$\mathbb{P}(\tau > t \mid \mathcal{F}_t) = \mathbb{E}\left[\mathbf{1}_{\{\tau > t\}} \mid \mathcal{F}_t\right] = \mathbf{1}_{\{\tau > t\}}, \qquad t \in \mathbb{R}_+.$$

In the sequel we will work in a framework where τ is not a stopping time with respect to $(\mathcal{F}_t)_{t \in \mathbb{R}_+}$, and we will assume that $\mathbb{P}(\tau > t \mid \mathcal{F}_t)$ takes the form

$$\mathbb{P}(\tau > t \mid \mathcal{F}_t) = \exp\left(-\int_0^t \lambda_u du\right), \qquad (9.4)$$

by analogy with (9.1).

This condition holds in particular when $(\mathcal{F}_t)_{t \in \mathbb{R}_+}$ is the filtration generated by a process $(X_t)_{t \in \mathbb{R}_+}$, λ_u has the form $\lambda_u = h(X_u)$, and τ is given by

$$\tau = \inf \left\{ t \in \mathbb{R}_+ \ : \ \int_0^t \lambda_u du \geq L \right\},$$

cf. [Lando (1998)].

We let $(\mathcal{G}_t)_{t \in \mathbb{R}_+}$ be the filtration defined by

$$\mathcal{G}_t = \mathcal{F}_t \vee \sigma(\{\tau < u\} \ : \ 0 \leq u \leq t), \qquad t \in \mathbb{R}_+,$$

i.e. $\mathcal{G}_t$ contains the additional information on whether default at time τ has occurred or not before time t.

Taking $F = 1$ in the next Lemma 9.1 shows that the probability of survival up to time T, given information known up to time t, is

$$\mathbb{P}(\tau > T \mid \mathcal{G}_t) = \mathbb{E}\left[\mathbf{1}_{\{\tau > T\}} \mid \mathcal{G}_t\right]$$

$$= \mathbf{1}_{\{\tau > t\}} \mathbb{E}\left[\exp\left(-\int_t^T \lambda_u du\right) \Big| \mathcal{F}_t\right], \qquad 0 \leq t \leq T. \quad (9.5)$$

Lemma 9.1. *([Guo et al. (2007)]) For any $\mathcal{F}_T$-measurable integrable random variable F we have*

$$\mathbb{E}\left[F\mathbf{1}_{\{\tau > T\}} \mid \mathcal{G}_t\right] = \mathbf{1}_{\{\tau > t\}} \mathbb{E}\left[F \exp\left(-\int_t^T \lambda_u du\right) \Big| \mathcal{F}_t\right].$$

Proof. By (9.4) we have

$$\frac{\mathbb{P}(\tau > T \mid \mathcal{F}_T)}{\mathbb{P}(\tau > t \mid \mathcal{F}_t)} = \frac{\exp\left(-\int_0^T \lambda_u du\right)}{\exp\left(-\int_0^t \lambda_u du\right)} = \exp\left(-\int_t^T \lambda_u du\right),$$

hence

$$\mathbf{1}_{\{\tau > t\}} \mathbb{E}\left[F\frac{\mathbb{P}(\tau > T \mid \mathcal{F}_T)}{\mathbb{P}(\tau > t \mid \mathcal{F}_t)} \Big| \mathcal{F}_t\right] = \frac{\mathbf{1}_{\{\tau > t\}}}{\mathbb{P}(\tau > t \mid \mathcal{F}_t)} \mathbb{E}\left[F\mathbb{E}[\mathbf{1}_{\{\tau > T\}} \mid \mathcal{F}_T] \Big| \mathcal{F}_t\right]$$

$$= \frac{\mathbf{1}_{\{\tau > t\}}}{\mathbb{P}(\tau > t \mid \mathcal{F}_t)} \mathbb{E}\left[\mathbb{E}[F\mathbf{1}_{\{\tau > T\}} \mid \mathcal{F}_T] \Big| \mathcal{F}_t\right]$$

$$= 1_{\{\tau>t\}} \frac{\mathbb{E}\left[F 1_{\{\tau>T\}} \middle| \mathcal{F}_t\right]}{\mathbb{P}(\tau > t \mid \mathcal{F}_t)}$$

$$= 1_{\{\tau>t\}} \mathbb{E}\left[F 1_{\{\tau>T\}} \middle| \mathcal{F}_t \vee \{\tau > t\}\right]$$

$$= 1_{\{\tau>t\}} \mathbb{E}\left[F 1_{\{\tau>T\}} \middle| \mathcal{G}_t\right], \qquad 0 \le t \le T. \qquad \square$$

The process λ_t can be chosen among the classical mean-reverting diffusion processes, including jump-diffusion processes.

The computation of $\mathbb{P}(\tau > T \mid \mathcal{G}_t)$ according to (9.5) is then similar to bond pricing. In addition, HJM type default intensities can also be introduced to compute

$$\mathbb{P}(\tau > T \mid \mathcal{G}_t) = \mathbb{E}\left[\exp\left(-\int_t^T \lambda(t,u)du\right) \middle| \mathcal{F}_t\right],$$

in which the process $\lambda(t,T)$ can be modeled in the HJM framework of Chapter 6.

9.3 Defaultable Bonds

The price of a bond with maturity T and (random) default time τ is given by

$$P(t,T) = \mathbb{E}\left[1_{\{\tau>T\}} \exp\left(-\int_t^T r_u du\right) \middle| \mathcal{G}_t\right]$$

$$+ \mathbb{E}\left[\xi 1_{\{\tau\le T\}} \exp\left(-\int_t^T r_u du\right) \middle| \mathcal{G}_t\right], \qquad 0 \le t \le T,$$

where $\xi \in [0,1]$ is the (possibly random) recovery rate.

Taking $F = \exp\left(-\int_t^T r_u du\right)$ in Lemma 9.1, we get

$$\mathbb{E}\left[1_{\{\tau>T\}} \exp\left(-\int_t^T r_u du\right) \middle| \mathcal{G}_t\right] = 1_{\{\tau>t\}} \mathbb{E}\left[\exp\left(-\int_t^T (r_u + \lambda_u)du\right) \middle| \mathcal{F}_t\right],$$

cf. e.g. [Lando (1998)], [Guo *et al.* (2007)], [Duffie and Singleton (2003)], hence

$$P(t,T) = 1_{\{\tau>t\}} \mathbb{E}\left[\exp\left(-\int_t^T (r_u + \lambda_u)du\right) \middle| \mathcal{F}_t\right]$$

$$+ \, \mathbb{E} \left[\xi \mathbf{1}_{\{\tau \leq T\}} \exp \left(- \int_t^T r_u du \right) \Big| \mathcal{G}_t \right], \qquad 0 \leq t \leq T.$$

In the case of complete default we have $\xi = 0$ (zero-recovery) and

$$P(t,T) = \mathbf{1}_{\{\tau > t\}} \, \mathbb{E} \left[\exp \left(- \int_t^T (r_s + \lambda_s) ds \right) \Big| \mathcal{F}_t \right], \qquad 0 \leq t \leq T. \quad (9.6)$$

The interest in this expression yields in the fact that the effect of the presence of a default time τ is to decrease the bond price that can be viewed as an increase λ_u in the short rate.

This treatment of default risk has some similarity with that of coupon bonds which can be priced as

$$P(t,T) = e^{c(T-t)} \, \mathbb{E} \left[\exp \left(- \int_t^T r_s ds \right) \Big| \mathcal{G}_t \right],$$

where $c > 0$ is a continuous-time coupon rate.

Finally, from Proposition 7.1 the bond price (9.6) can also be expressed under the forward measure $\tilde{\mathbb{P}}$ with maturity T, as

$$P(t,T) = \mathbf{1}_{\{\tau > t\}} \, \mathbb{E} \left[\exp \left(- \int_t^T (r_s + \lambda_s) ds \right) \Big| \mathcal{F}_t \right]$$

$$= \mathbf{1}_{\{\tau > t\}} \, \mathbb{E} \left[\exp \left(- \int_t^T r_s ds \right) \Big| \mathcal{F}_t \right] \mathbb{E}_{\tilde{\mathbb{P}}} \left[\exp \left(- \int_t^T \lambda_s ds \right) \Big| \mathcal{F}_t \right]$$

$$= \mathbf{1}_{\{\tau > t\}} \, \mathbb{E} \left[\exp \left(- \int_t^T r_s ds \right) \Big| \mathcal{F}_t \right] Q(t,T),$$

where

$$Q(t,T) = \mathbb{E}_{\tilde{\mathbb{P}}} \left[\exp \left(- \int_t^T \lambda_s ds \right) \Big| \mathcal{F}_t \right]$$

denotes the survival probability under the forward measure $\tilde{\mathbb{P}}$, cf. [Chen and Huang (2001)], [Chen *et al.* (2008)].

9.4   Credit Default Swaps

We work with a tenor structure $\{t = T_i < \cdots < T_j = T\}$.

A Credit Default Swap (CDS) is a contract consisting in

- **a premium leg:** the buyer is purchasing protection at time t against default at time T_k, $k = i+1, \ldots, j$, and has to make a fixed payment S_t at times $T_{i+1}, \ldots, T_j$ between t and T in compensation.

The discounted value at time t of the premium leg is

$$
\begin{aligned}
V(t,T) &= \mathbb{E}\left[\sum_{k=i}^{j-1} S_t \delta_k \mathbf{1}_{\{\tau > T_{k+1}\}} \exp\left(-\int_t^{T_{k+1}} r_s ds\right) \bigg| \mathcal{G}_t\right] \\
&= \sum_{k=i}^{j-1} S_t \delta_k \, \mathbb{E}\left[\mathbf{1}_{\{\tau > T_{k+1}\}} \exp\left(-\int_t^{T_{k+1}} r_s ds\right) \bigg| \mathcal{G}_t\right] \\
&= S_t \sum_{k=i}^{j-1} \delta_k P(t, T_{k+1}) \\
&= S_t P(t, T_i, T_j),
\end{aligned}
$$

where $\delta_k = T_{k+1} - T_k$, $P(t, T_i, T_j)$ is the annuity numeraire (10.6), and

$$
P(t, T_k) = \mathbf{1}_{\{\tau > t\}} \, \mathbb{E}\left[\exp\left(-\int_t^{T_k} (r_s + \lambda_s) ds\right) \bigg| \mathcal{F}_t\right], \qquad 0 \le t \le T_k,
$$

is the defaultable bond price with maturity T_k, $k = i, \ldots, j-1$. For simplicity we have ignored a possible accrual interest term over the time period $[T_k, \tau]$ when $\tau \in [T_k, T_{k+1}]$ in the above value of the premium leg.

- **a protection leg:** the seller or issuer of the contract makes a payment $1 - \xi_{k+1}$ to the buyer in case default occurs at time T_{k+1}, $k = i, \ldots, j-1$.

The value at time t of the protection leg is

$$
\mathbb{E}\left[\sum_{k=i}^{j-1} \mathbf{1}_{(T_k, T_{k+1}]}(\tau)(1 - \xi_{k+1}) \exp\left(-\int_t^{T_{k+1}} r_s ds\right) \bigg| \mathcal{G}_t\right],
$$

where ξ_{k+1} is the recovery rate associated with the maturity T_{k+1}, $k = i, \ldots, j-1$.

In the case of a non-random recovery rate ξ_k the value of the protection leg becomes

$$
\sum_{k=i}^{j-1} (1 - \xi_{k+1}) \, \mathbb{E}\left[\mathbf{1}_{(T_k, T_{k+1}]}(\tau) \exp\left(-\int_t^{T_{k+1}} r_s ds\right) \bigg| \mathcal{G}_t\right].
$$

The spread S_t is computed by equating the values of the protection and premium legs, i.e. from the relation

$$V(t,T) = S_t P(t,T_i,T_j)$$

$$= \mathbb{E}\left[\sum_{k=i}^{j-1} \mathbf{1}_{(T_k,T_{k+1}]}(\tau)(1-\xi_{k+1}) \exp\left(-\int_t^{T_{k+1}} r_s ds\right) \Big| \mathcal{G}_t\right],$$

which yields

$$S_t = \frac{1}{P(t,T_i,T_j)} \sum_{k=i}^{j-1} \mathbb{E}\left[\mathbf{1}_{(T_k,T_{k+1}]}(\tau)(1-\xi_{k+1}) \exp\left(-\int_t^{T_{k+1}} r_s ds\right) \Big| \mathcal{G}_t\right].$$

In the case of a constant recovery rate ξ we find

$$S_t = \frac{1-\xi}{P(t,T_i,T_j)} \sum_{k=i}^{j-1} \mathbb{E}\left[\mathbf{1}_{(T_k,T_{k+1}]}(\tau) \exp\left(-\int_t^{T_{k+1}} r_s ds\right) \Big| \mathcal{G}_t\right],$$

and if τ is constrained to take values in the tenor structure $\{T_i,\ldots,T_j\}$ we get

$$S_t = \frac{1-\xi}{P(t,T_i,T_j)} \mathbb{E}\left[\mathbf{1}_{[t,T]}(\tau) \exp\left(-\int_t^{\tau} r_s ds\right) \Big| \mathcal{G}_t\right].$$

9.5   Exercises

Problem 9.1. Defaultable bonds. Consider a (random) default time τ with law

$$\mathbb{P}(\tau > t \mid \mathcal{F}_t) = \exp\left(-\int_0^t \lambda_u du\right),$$

where λ_t is a (random) default rate process which is adapted to the filtration $(\mathcal{F}_t)_{t\in\mathbb{R}_+}$. Recall that the probability of survival up to time T, given information known up to time t, is given by

$$\mathbb{P}(\tau > T \mid \mathcal{G}_t) = \mathbf{1}_{\{\tau>t\}} E\left[\exp\left(-\int_t^T \lambda_u du\right) \Big| \mathcal{F}_t\right],$$

where $\mathcal{G}_t = \mathcal{F}_t \vee \sigma(\{\tau < u\} : 0 \leq u \leq t)$, $t \in \mathbb{R}_+$, is the filtration defined by adding the default time information to the history $(\mathcal{F}_t)_{t\in\mathbb{R}_+}$. In this framework, the price $P(t,T)$ of defaultable bond with maturity T, short term interest rate r_t and (random) default time τ is given by

$$P(t,T) = E\left[\mathbf{1}_{\{\tau>T\}} \exp\left(-\int_t^T r_u du\right) \Big| \mathcal{G}_t\right] \tag{9.7}$$

$$= \mathbf{1}_{\{\tau > t\}} E\left[\exp\left(-\int_t^T (r_u + \lambda_u)du\right)\bigg|\mathcal{F}_t\right].$$

In the sequel we assume that the processes $(r_t)_{t \in \mathbb{R}_+}$ and $(\lambda_t)_{t \in \mathbb{R}_+}$ are modeled according to the Vasicek processes

$$\begin{cases} dr_t = -ar_t dt + \sigma dB_t^1, \\[2mm] d\lambda_t = -b\lambda_t dt + \eta dB_t^2, \end{cases}$$

where $(B_t^1)_{t \in \mathbb{R}_+}$ and $(B_t^2)_{t \in \mathbb{R}_+}$ are two standard $\mathcal{F}_t$-Brownian motions with correlation $\rho \in [-1, 1]$, and $dB_t^1 dB_t^2 = \rho dt$.

(1) Give a justification for the fact that

$$E\left[\exp\left(-\int_t^T (r_u + \lambda_u)du\right)\bigg|\mathcal{F}_t\right]$$

can be written as a function $F(t, r_t, \lambda_t)$ of t, r_t and λ_t, $t \in [0, T]$.

(2) Show that

$$t \longmapsto \exp\left(-\int_0^t (r_s + \lambda_s)ds\right) E\left[\exp\left(-\int_t^T (r_u + \lambda_u)du\right)\bigg|\mathcal{F}_t\right]$$

is an $\mathcal{F}_t$-martingale under $\mathbb{P}$.

(3) Use the Itô formula with two variables to derive a PDE on $\mathbb{R}^2$ for the function $F(t, x, y)$.

(4) Show that we have

$$\int_t^T r_s ds = C(a, t, T)r_t + \sigma \int_t^T C(a, s, T)dB_s^1,$$

and

$$\int_t^T \lambda_s ds = C(b, t, T)\lambda_t + \eta \int_t^T C(b, s, T)dB_s^2,$$

where

$$C(a, t, T) = -\frac{1}{a}(e^{-a(T-t)} - 1).$$

(5) Show that the random variable

$$\int_t^T r_s ds + \int_t^T \lambda_s ds$$

is Gaussian and compute its mean

$$\mathbb{E}\left[\int_t^T r_s ds + \int_t^T \lambda_s ds\bigg|\mathcal{F}_t\right]$$

and variance

$$\mathrm{Var}\left[\int_t^T r_s ds + \int_t^T \lambda_s ds \Big| \mathcal{F}_t\right],$$

conditionally to $\mathcal{F}_t$.

(6) Compute $P(t,T)$ from its expression (9.7) as a conditional expectation.

(7) Show that the solution $F(t,x,y)$ to the 2-dimensional PDE of Question (3) is

$$F(t,x,y) = \exp\left(-C(a,t,T)x - C(b,t,T)y\right)$$

$$\times \exp\left(\frac{\sigma^2}{2}\int_t^T C^2(a,s,T)ds + \frac{\eta^2}{2}\int_t^T C^2(b,s,T)ds\right)$$

$$\times \exp\left(\rho\sigma\eta\int_t^T C(a,s,T)C(b,s,T)ds\right).$$

(8) Show that the defaultable bond price $P(t,T)$ can also be written as

$$P(t,T) = e^{U(t,T)}\mathbb{P}(\tau > T \mid \mathcal{G}_t)\,\mathbb{E}\left[\exp\left(-\int_t^T r_s ds\right)\Big|\mathcal{F}_t\right],$$

where

$$U(t,T) = \rho\frac{\sigma\eta}{ab}\left(T - t - C(a,t,T) - C(b,t,T) + C(a+b,t,T)\right).$$

(9) By partial differentiation of $\log P(t,T)$ with respect to T, compute the corresponding instantaneous short rate $f(t,T) = -\frac{\partial \log P(t,T)}{\partial T}$.

(10) Show that $\mathbb{P}(\tau > T \mid \mathcal{G}_t)$ can be written using an HJM type default rate as

$$\mathbb{P}(\tau > T \mid \mathcal{G}_t) = \mathbf{1}_{\{\tau > t\}}\exp\left(-\int_t^T f_2(t,u)du\right),$$

where

$$f_2(t,u) = \lambda_t e^{-b(u-t)} - \frac{\eta^2}{2}C^2(b,t,u).$$

(11) Show how the result of Question (8) can be simplified when $(B_t^1)_{t\in\mathbb{R}_+}$ and $(B_t^2)_{t\in\mathbb{R}_+}$ are independent.

Chapter 10

Pricing of Caps and Swaptions on the LIBOR

In this chapter we consider the pricing of caps and swaptions using forward measures on the London Interbank Offered Rates (LIBOR), in which forward rates are defined using different compounding conventions. We also introduce the swap rates to be used in the next Chapter 11 for the Brace-Gatarek-Musiela (BGM) model.

10.1 Pricing of Caplets and Caps

Recall, cf. Chapter 7, that the caplet on the spot rate $f(T, T, S)$ with strike κ is a contract with payoff

$$(f(T, T, S) - \kappa)^+,$$

priced at time $t \in [0, T]$ under the forward measure as

$$\mathbb{E}\left[e^{-\int_t^S r_s ds}(f(T, T, S) - \kappa)^+ \Big| \mathcal{F}_t\right] = P(t, S)\,\mathbb{E}_S\left[(f(T, T, S) - \kappa)^+ \mid \mathcal{F}_t\right],$$

$$(10.1)$$

where $\mathbb{E}_S$ denotes the expectation under the forward measure $\mathbb{P}_S$ with maturity S and density

$$\frac{d\mathbb{P}_S}{d\mathbb{P}} = \frac{1}{P(0, S)} e^{-\int_0^S r_s ds},$$

i.e.

$$\frac{d\mathbb{P}_{S|\mathcal{F}_t}}{d\mathbb{P}_{|\mathcal{F}_t}} = \frac{e^{-\int_t^S r_s ds}}{P(t, S)}, \qquad \text{or} \qquad \mathbb{E}_{\mathbb{P}}\left[\frac{d\mathbb{P}_S}{d\mathbb{P}} \Big| \mathcal{F}_t\right] = \frac{P(t, S)}{P(0, S)} e^{-\int_0^t r_s ds},$$

$0 \le t \le S.$

In practice, the maturity dates are arranged according to a discrete *tenor structure*

$$\{0 = T_0 < T_1 < T_2 < \cdots < T_n\}.$$

An example of data used for the forward interest rate curve of Figure 5.1 is given in the table of Figure 10.1, which contains the values of $(T_1, T_2, \ldots, T_{23})$ and of $\{f(t, t+T_i, t+T_i+\delta)\}_{i=1,\ldots,23}$, with $t = 07/05/2003$ and $\delta = $ six months.

TimeSerieNb	505
AsOfDate	7-mai-03
2D	2,55
1W	2,53
1M	2,56
2M	2,52
3M	2,48
1Y	2,34
2Y	2,49
3Y	2,79
4Y	3,07
5Y	3,31
6Y	3,52
7Y	3,71
8Y	3,88
9Y	4,02
10Y	4,14
11Y	4,23
12Y	4,33
13Y	4,4
14Y	4,47
15Y	4,54
20Y	4,74
25Y	4,83
30Y	4,86

Fig. 10.1   Forward rates according to a tenor structure.

More generally, instead of caplets one can consider caps that are relative to a given tenor structure $\{T_i, \ldots, T_j\}$, $1 \le i < j \le n$, with discounted payoff

$$\sum_{k=i}^{j-1}(T_{k+1} - T_k)e^{-\int_t^{T_{k+1}} r_s ds}(f(T_k, T_k, T_{k+1}) - \kappa)^+.$$

Pricing formulas for caps are easily deduced from analog formulas for caplets, since the payoff of a cap can be decomposed into a sum of caplet payoffs. Thus the price of a cap at time $t \in [0, T_i]$ is given by

$$\mathbb{E}\left[\sum_{k=i}^{j-1}(T_{k+1} - T_k)e^{-\int_t^{T_{k+1}} r_s ds}(f(T_k, T_k, T_{k+1}) - \kappa)^+ \Big| \mathcal{F}_t\right]$$

$$= \sum_{k=i}^{j-1}(T_{k+1} - T_k)\,\mathbb{E}\left[e^{-\int_t^{T_{k+1}} r_s ds}(f(T_k, T_k, T_{k+1}) - \kappa)^+ \Big| \mathcal{F}_t\right]$$

$$= \sum_{k=i}^{j-1} (T_{k+1} - T_k) P(t, T_{k+1}) \, \mathbb{E}_{k+1} \left[(f(T_k, T_k, T_{k+1}) - \kappa)^+ \Big| \mathcal{F}_t \right],$$

where $\mathbb{E}_{k+1}$ denotes the expectation under the forward measure $\mathbb{P}_{k+1}$ defined as

$$\frac{d\mathbb{P}_{k+1}}{d\mathbb{P}} = \frac{1}{P(0, T_{k+1})} e^{-\int_0^{T_{k+1}} r_s ds}, \qquad k = 0, \ldots, n-1,$$

cf. Definition 7.1.

10.2 Forward Rate Measure and Tenor Structure

In this section we review the construction of the multiple forward measures $\mathbb{P}_i$, $i = 1, \ldots, n$. Recall that the absence of arbitrage condition states that

$$t \longmapsto e^{-\int_0^t r_s ds} P(t, T_i), \qquad 0 \le t \le T_i, \quad i = 1, \ldots, n,$$

is an $\mathcal{F}_t$-martingale under $\mathbb{P}$.

Definition 10.1. *The probability measure $\mathbb{P}_i$ is defined as*

$$\frac{d\mathbb{P}_i}{d\mathbb{P}} = \frac{1}{P(0, T_i)} e^{-\int_0^{T_i} r_s ds}, \qquad i = 1, \ldots, n.$$

Note that for $i = 1, \ldots, n$, we have

$$\mathbb{E}\left[\frac{d\mathbb{P}_i}{d\mathbb{P}} \Big| \mathcal{F}_t \right] = \frac{1}{P(0, T_i)} \mathbb{E}\left[e^{-\int_0^{T_i} r_s ds} \Big| \mathcal{F}_t \right]$$

$$= \frac{e^{-\int_0^t r_s ds}}{P(0, T_i)} \mathbb{E}\left[e^{-\int_t^{T_i} r_s ds} \Big| \mathcal{F}_t \right]$$

$$= \frac{P(t, T_i)}{P(0, T_i)} e^{-\int_0^t r_s ds}, \qquad 0 \le t \le T_i.$$

Moreover, for all $i = 1, \ldots, n$ we have

$$\frac{d\mathbb{P}_{i|\mathcal{F}_t}}{d\mathbb{P}_{|\mathcal{F}_t}} = \frac{e^{-\int_t^{T_i} r_s ds}}{P(t, T_i)}, \qquad 0 \le t \le T_i. \tag{10.2}$$

Indeed, for all bounded and $\mathcal{F}_t$-measurable random variables G,

$$\mathbb{E}\left[GFe^{-\int_t^{T_i} r_s ds} \right] = P(0, T_i) \mathbb{E}_i \left[Ge^{\int_0^t r_s ds} F \right]$$

$$= P(0, T_i) \mathbb{E}_i \left[Ge^{\int_0^t r_s ds} \mathbb{E}_i[F \mid \mathcal{F}_t] \right]$$

$$= \mathbb{E}\left[Ge^{-\int_t^{T_i} r_s ds} \mathbb{E}_i[F \mid \mathcal{F}_t] \right]$$

$$= P(t, T_i) \mathbb{E}\left[G \mathbb{E}_i[F \mid \mathcal{F}_t] \right],$$

hence

$$\mathbb{E}\left[Fe^{-\int_t^{T_i} r_s ds}\Big|\mathcal{F}_t\right] = P(t, T_i)\,\mathbb{E}_i[F \mid \mathcal{F}_t], \qquad 0 \le t \le T_i,$$

for all integrable random variables F, which shows (10.2).

From now on we assume as in (7.7) that the dynamics of the bond price $P(t, T_i)$ is given by

$$\frac{dP(t, T_i)}{P(t, T_i)} = r_t dt + \zeta_i(t) dB_t, \qquad (10.3)$$

for $i = 1, \ldots, n$. The next statement parallels Proposition 7.3.

Proposition 10.1. *For $i = 1, \ldots, n$, let*

$$B_t^i := B_t - \int_0^t \zeta_i(s) ds, \qquad 0 \le t \le T_i, \qquad (10.4)$$

then $(B_t^i)_{t \in [0, T_i]}$ is a standard Brownian motion under $\mathbb{P}_i$.

Proof. Letting

$$\Phi_i(t) = \mathbb{E}\left[\frac{d\mathbb{P}_i}{d\mathbb{P}}\Big|\mathcal{F}_t\right] = \frac{P(t, T_i)}{P(0, T_i)} e^{-\int_0^t r_s ds}, \qquad 0 \le t \le T_i,$$

we have $d\Phi_i(t) = \Phi_i(t)\zeta_i(t) dB_t$ and $\Phi_i(0) = 1$, hence

$$\mathbb{E}\left[\frac{d\mathbb{P}_i}{d\mathbb{P}}\Big|\mathcal{F}_{T_i}\right] = \Phi_i(T_i) = \exp\left(\int_0^{T_i} \zeta_i(t) dB_t - \frac{1}{2}\int_0^{T_i} |\zeta_i(t)|^2 dt\right), \quad 0 \le t \le T_i,$$

and by the Girsanov Theorem 2.1,

$$B_t - \int_0^t \frac{1}{\Phi_i(s)} d\langle \Phi_i, B\rangle_s = B_t - \int_0^t \zeta_i(s) ds, \qquad 0 \le t \le T_i,$$

is a continuous martingale under $\mathbb{P}_i$. $\square$

We also have the relation

$$dB_t^j = dB_t^i - (\zeta_j(t) - \zeta_i(t)) dt.$$

Recall that the expectation under $\mathbb{P}_i$ is denoted by $\mathbb{E}_i$. The next result is related to Proposition 7.2.

Proposition 10.2. *For all $1 \le i, j \le n$ we have*

$$\mathbb{E}_i\left[\frac{d\mathbb{P}_j}{d\mathbb{P}_i}\Big|\mathcal{F}_t\right] = \frac{P(0, T_i)}{P(0, T_j)} \frac{P(t, T_j)}{P(t, T_i)} \qquad 0 \le t \le T_i \wedge T_j, \qquad (10.5)$$

and in particular the process

$$t \longmapsto \frac{P(t, T_j)}{P(t, T_i)}, \qquad 0 \le t \le T_i \wedge T_j,$$

is an $\mathcal{F}_t$-martingale under $\mathbb{P}_i$, $1 \le i, j \le n$.

Proof. For all bounded and $\mathcal{F}_t$-measurable random variables F we have[1]

$$\mathbb{E}_i\left[F\frac{d\mathbb{P}_j}{d\mathbb{P}_i}\right] = \mathbb{E}\left[F\frac{d\mathbb{P}_j}{d\mathbb{P}}\right]$$

$$= \frac{1}{P(0,T_j)}\,\mathbb{E}\left[Fe^{-\int_0^{T_j} r_\tau d\tau}\right]$$

$$= \frac{1}{P(0,T_j)}\,\mathbb{E}\left[Fe^{-\int_0^t r_\tau d\tau}P(t,T_j)\right]$$

$$= \frac{1}{P(0,T_j)}\,\mathbb{E}\left[Fe^{-\int_0^{T_i} r_\tau d\tau}\frac{P(t,T_j)}{P(t,T_i)}\right]$$

$$= \mathbb{E}_i\left[F\frac{P(0,T_i)}{P(0,T_j)}\frac{P(t,T_j)}{P(t,T_i)}\right],$$

which shows (10.5). □

By Itô's calculus we have from (10.3), for any $1 \le i < j \le n$,

$$d\left(\frac{P(t,T_j)}{P(t,T_i)}\right) = \frac{P(t,T_j)}{P(t,T_i)}(\zeta_j(t) - \zeta_i(t))(dB_t - \zeta_i(t)dt)$$

$$= \frac{P(t,T_j)}{P(t,T_i)}(\zeta_j(t) - \zeta_i(t))dB_t^i,$$

which, from Proposition 10.1, recovers the second part of Proposition 10.2, i.e. the martingale property of $P(t,T_j)/P(t,T_i)$.

We close this section with the definition of the *annuity numeraire*

$$P(t,T_i,T_j) = \sum_{k=i}^{j-1}(T_{k+1} - T_k)P(t,T_{k+1}), \qquad 0 \le t \le T_i, \qquad (10.6)$$

$1 \le i < j \le n$, which can be viewed as the value at time t of a sequence of \$1 payments received at the times $T_{i+1},\ldots,T_j$, weighted by the interval lengths $\delta_k = T_{k+1} - T_k$, $k = i,\ldots,j-1$.

The annuity numeraire satisfies the following martingale property.

Proposition 10.3. *The discounted annuity numeraire*

$$t \longmapsto e^{-\int_0^t r_s ds}P(t,T_i,T_j), \qquad 0 \le t \le T_i,$$

is a martingale under $\mathbb{P}$, $1 \le i < j \le n$.

---

[1]We will repeatedly use the characterization $X = \mathbb{E}[F|\mathcal{F}_t] \Leftrightarrow \mathbb{E}[GX] = \mathbb{E}[GF]$ for all bounded and $\mathcal{F}_t$-measurable random variable G, cf. (12.4) in the Appendix.

Proof. This result can be recovered by linearity and the fact that $t \longmapsto$ $e^{-\int_0^t r_s ds} P(t, T_k)$ is a martingale for all $k = i, \ldots, j$. Alternatively, by standard arguments, given that $\delta_k = T_{k+1} - T_k$, $k = i, \ldots, j-1$, we have

$$\mathbb{E}\left[e^{-\int_0^T r_s ds} P(T, T_i, T_j) \middle| \mathcal{F}_t\right] = \sum_{k=i}^{j-1} \delta_k \, \mathbb{E}\left[e^{-\int_0^T r_s ds} P(T, T_{k+1}) \middle| \mathcal{F}_t\right]$$

$$= \sum_{k=i}^{j-1} \delta_k \, \mathbb{E}\left[e^{-\int_0^T r_s ds} \mathbb{E}\left[e^{-\int_T^{T_{k+1}} r_s ds} \middle| \mathcal{F}_T\right] \middle| \mathcal{F}_t\right]$$

$$= \sum_{k=i}^{j-1} \delta_k \, \mathbb{E}\left[e^{-\int_0^T r_s ds} \mathbb{E}\left[e^{-\int_T^{T_{k+1}} r_s ds} \middle| \mathcal{F}_T\right] \middle| \mathcal{F}_t\right]$$

$$= \sum_{k=i}^{j-1} \delta_k \, \mathbb{E}\left[\mathbb{E}\left[e^{-\int_0^T r_s ds} e^{-\int_T^{T_{k+1}} r_s ds} \middle| \mathcal{F}_T\right] \middle| \mathcal{F}_t\right]$$

$$= \sum_{k=i}^{j-1} \delta_k \, \mathbb{E}\left[\mathbb{E}\left[e^{-\int_0^{T_{k+1}} r_s ds} \middle| \mathcal{F}_T\right] \middle| \mathcal{F}_t\right]$$

$$= \sum_{k=i}^{j-1} \delta_k \, \mathbb{E}\left[e^{-\int_0^{T_{k+1}} r_s ds} \middle| \mathcal{F}_t\right]$$

$$= \sum_{k=i}^{j-1} \delta_k e^{-\int_0^t r_s ds} P(t, T_{k+1})$$

$$= e^{-\int_0^t r_s ds} P(t, T_i, T_j),$$

for $0 \leq t \leq T \leq T_i$. □

In particular we have

$$P(t, T_i, T_{i+1}) = (T_{i+1} - T_i) P(t, T_{i+1}),$$

$0 \leq t \leq T_i$, $1 \leq i < n$.

Note that

$$d\left(e^{-\int_0^t r_s ds} P(t, T_i, T_j)\right) = \sum_{k=i}^{j-1} d\left(e^{-\int_0^t r_s ds} P(t, T_{k+1})\right)$$

$$= e^{-\int_0^t r_s ds} \sum_{k=i}^{j-1} \zeta_{k+1}(t) P(t, T_{k+1}) dB_t$$

$$= e^{-\int_0^t r_s ds} P(t, T_i, T_j) \sum_{k=i}^{j-1} v_{k+1}^{i,j}(t) \zeta_{k+1}(t) dB_t,$$

where

$$v_k^{i,j}(t) := \frac{P(t,T_k)}{P(t,T_i,T_j)}, \qquad 0 \le t \le T_i,$$

$1 \le i < j \le n$, $1 \le k \le n$, which recovers the result of Proposition 10.3.

10.3 Swaps and Swaptions

An interest rate swap makes it possible to exchange a variable forward rate $f(t,T,S)$ against a fixed rate κ. Such an exchange will generate a cash flow valued *at time t* as

$$\sum_{k=i}^{j-1} (T_{k+1} - T_k) P(t, T_{k+1}) (f(t, T_k, T_{k+1}) - \kappa).$$

The swap rate $S(t, T_i, T_j)$ is by definition the fair value of κ that cancels this cash flow:

$$\sum_{k=i}^{j-1} (T_{k+1} - T_k) P(t, T_{k+1}) (f(t, T_k, T_{k+1}) - S(t, T_i, T_j)) = 0 \qquad (10.7)$$

is called the swap rate $S(t, T_i, T_j)$, and it satisfies

$$S(t, T_i, T_j) = \frac{1}{P(t, T_i, T_j)} \sum_{k=i}^{j-1} (T_{k+1} - T_k) P(t, T_{k+1}) f(t, T_k, T_{k+1}), \quad (10.8)$$

where $P(t, T_i, T_j)$ is the annuity numeraire defined in (10.6).

In particular, when $j = i + 1$ we get

$$S(t, T_i, T_{i+1}) = f(t, T_i, T_{i+1}),$$

i.e. in this case the forward rate and the swap rate coincide.

A swaption is a contract to protect oneself against a risk based on an interest rate swap, and has payoff

$$\left(\sum_{k=i}^{j-1} (T_{k+1} - T_k) e^{- \int_{T_i}^{T_{k+1}} r_s ds} f(T_i, T_k, T_{k+1}) - \kappa \right)^+.$$

This swaption can be priced at time $t \in [0, T_i]$ as

$$\mathbb{E}\left[e^{- \int_t^{T_i} r_s ds} \left(\sum_{k=i}^{j-1} (T_{k+1} - T_k) e^{- \int_{T_i}^{T_{k+1}} r_s ds} (f(T_i, T_k, T_{k+1}) - \kappa) \right)^+ \Bigg| \mathcal{F}_t \right].$$

$$(10.9)$$

Unlike in the case of caps, the sum can not be taken out of the positive part. Nevertheless, the price of the swaption can be bounded as follows:

$$
\mathbb{E}\left[e^{-\int_t^{T_i} r_s ds} \left(\sum_{k=i}^{j-1} (T_{k+1} - T_k) e^{-\int_{T_i}^{T_{k+1}} r_s ds} (f(T_i, T_k, T_{k+1}) - \kappa) \right)^+ \bigg| \mathcal{F}_t \right]
$$

$$
\leq \mathbb{E}\left[e^{-\int_t^{T_i} r_s ds} \sum_{k=i}^{j-1} (T_{k+1} - T_k) e^{-\int_{T_i}^{T_{k+1}} r_s ds} (f(T_i, T_k, T_{k+1}) - \kappa)^+ \bigg| \mathcal{F}_t \right]
$$

$$
= \sum_{k=i}^{j-1} (T_{k+1} - T_k) \mathbb{E}\left[e^{-\int_t^{T_{k+1}} r_s ds} (f(T_i, T_k, T_{k+1}) - \kappa)^+ \bigg| \mathcal{F}_t \right]
$$

$$
= \sum_{k=i}^{j-1} (T_{k+1} - T_k) P(t, T_{k+1}) \mathbb{E}_{k+1}\left[(f(T_i, T_k, T_{k+1}) - \kappa)^+ \big| \mathcal{F}_t \right].
$$

In the sequel and in practice the price (10.9) of the swaption will be evaluated as

$$
\mathbb{E}\left[e^{-\int_t^{T_i} r_s ds} \left(\sum_{k=i}^{j-1} (T_{k+1} - T_k) P(T_i, T_{k+1}) (f(T_i, T_k, T_{k+1}) - \kappa) \right)^+ \bigg| \mathcal{F}_t \right],
$$

$$(10.10)$$

meaning that we approximate the discount factor $e^{-\int_{T_i}^{T_{k+1}} r_s ds}$ by its conditional expectation $P(T_i, T_{k+1})$ given $\mathcal{F}_{T_i}$. The use of (10.10) instead of (10.9) will be essential in computing the price of swaptions in connection with the LIBOR, see (10.23) below.

Note that when $j = i + 1$, the swaption price (10.10) coincides with the price at time t of a caplet on $[T_i, T_{i+1}]$ up to a factor $\delta_i := T_{i+1} - T_i$ since

$$
\mathbb{E}\left[e^{-\int_t^{T_i} r_s ds} ((T_{i+1} - T_i) P(T_i, T_{i+1}) (f(T_i, T_i, T_{i+1}) - \kappa))^+ \bigg| \mathcal{F}_t \right]
$$

$$
= (T_{i+1} - T_i) \mathbb{E}\left[e^{-\int_t^{T_i} r_s ds} P(T_i, T_{i+1}) ((f(T_i, T_i, T_{i+1}) - \kappa))^+ \bigg| \mathcal{F}_t \right]
$$

$$
= \delta_i \mathbb{E}\left[e^{-\int_t^{T_i} r_s ds} \mathbb{E}\left[e^{-\int_{T_i}^{T_{i+1}} r_s ds} \bigg| \mathcal{F}_{T_i} \right] ((f(T_i, T_i, T_{i+1}) - \kappa))^+ \bigg| \mathcal{F}_t \right]
$$

$$
= \delta_i \mathbb{E}\left[\mathbb{E}\left[e^{-\int_t^{T_i} r_s ds} e^{-\int_{T_i}^{T_{i+1}} r_s ds} ((f(T_i, T_i, T_{i+1}) - \kappa))^+ \bigg| \mathcal{F}_{T_i} \right] \bigg| \mathcal{F}_t \right]
$$

$$
= \delta_i \mathbb{E}\left[e^{-\int_t^{T_{i+1}} r_s ds} (f(T_i, T_i, T_{i+1}) - \kappa)^+ \bigg| \mathcal{F}_t \right],
$$

$0 \leq t \leq T_i$.

10.4   The London InterBank Offered Rates (LIBOR) Model

Recall that the forward rate $f(t, T, S)$, $0 \leq t \leq T \leq S$, has been defined from the relation

$$P(t, T) - P(t, S) \exp\left((S - T)f(t, T, S)\right) = 0, \tag{10.11}$$

or

$$\exp\left((S - T)f(t, T, S)\right) = \frac{P(t, T)}{P(t, S)},$$

i.e.

$$f(t, T, S) = -\frac{\log P(t, S) - \log P(t, T)}{S - T}.$$

In order to compute swaption prices one prefers to use forward rates as defined on the London InterBank Offered Rates (LIBOR) market instead of the standard forward rates given by (10.11).

Just as in Section 5.1, a forward rate agreement at time t on the LIBOR market also gives its holder an interest rate $L(t, T, S)$ over the future time period $[T, S]$. However, instead of using exponential compounding of rates, the forward LIBOR $L(t, T, S)$ for a loan on $[T, S]$ is defined using linear compounding, i.e. by replacing (10.11) with the relation

$$1 + (S - T)L(t, T, S) = \frac{P(t, T)}{P(t, S)}.$$

Equivalently we have

$$P(t, T) - P(t, S) - P(t, S)(S - T)L(t, T, S) = 0,$$

$0 \leq t \leq T < S$, which yields the following definition.

Definition 10.2. *The forward LIBOR rate $L(t, T, S)$ at time t for a loan on $[T, S]$ is given by*

$$L(t, T, S) = \frac{1}{S - T}\left(\frac{P(t, T)}{P(t, S)} - 1\right), \qquad 0 \leq t \leq T < S.$$

A model for the dynamics of LIBOR rates (the BGM model) will be considered in Chapter 11.

Note that if $1 \leq i < j \leq n$ we have

$$P(t, T_j) = P(t, T_i) \prod_{k=i}^{j-1} \frac{1}{1 + (T_{k+1} - T_k)L(t, T_k, T_{k+1})}, \qquad 0 \leq t \leq T_i,$$

---

and if $1 \le j \le i \le n$,

$$P(t, T_j) = P(t, T_i) \prod_{k=j}^{i-1} (1 + (T_{k+1} - T_k)L(t, T_k, T_{k+1})), \qquad 0 \le t \le T_j.$$

The instantaneous forward rate $f(t, T)$ can be recovered from the forward instantaneous LIBOR rate

$$L(t, T) = \lim_{S \searrow T} L(t, T, S)$$

as

$$f(t, T) = -\frac{\partial}{\partial T} \log P(t, T)$$

$$= -\frac{1}{P(t, T)} \frac{\partial P}{\partial T}(t, T)$$

$$= -\frac{1}{P(t, T)} \lim_{S \searrow T} \frac{P(t, S) - P(t, T)}{S - T}$$

$$= -\lim_{S \searrow T} \frac{P(t, S) - P(t, T)}{(S - T)P(t, S)}$$

$$= \lim_{S \searrow T} L(t, T, S)$$

$$= L(t, T, T)$$

$$= L(t, T).$$

In this model the short term interest rate thus satisfies

$$r_t = L(t, t), \qquad t \in \mathbb{R}_+,$$

thus the LIBOR and standard instantaneous rates coincide.

Figure 10.2 presents a simulation of the simply compounded spot rates $t \longmapsto L(t, t, T)$, computed from previous sample graphs of Chapter 4 in the Vasicek model.

The forward curve $T \longmapsto L(0, T, T + \delta)$ is plotted in Figure 10.3 for $t = 0$, also using bond prices computed in the Vasiçek model.

10.5 Swap Rates on the LIBOR Market

From (10.7) the forward swap rate $S(t, T_i, T_j)$ on the LIBOR market satisfies

$$\sum_{k=i}^{j-1} (T_{k+1} - T_k)P(t, T_{k+1})(L(t, T_k, T_{k+1}) - S(t, T_i, T_j)) = 0.$$

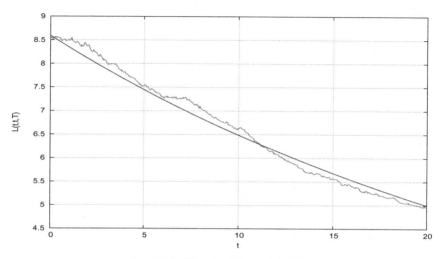

Fig. 10.2   Graph of $t \longmapsto L(t, t, T)$.

Fig. 10.3   Graph of $T \longmapsto L(0, T, T + \delta)$.

Proposition 10.4. *We have*

$$S(t, T_i, T_j) = \frac{P(t, T_i) - P(t, T_j)}{P(t, T_i, T_j)}, \qquad 0 \le t \le T_i, \quad 1 \le i < j \le n.$$

Proof.   By definition of the forward LIBOR $L(t, T, S)$ we have

$$P(t, T_k) - P(t, T_{k+1}) - (T_{k+1} - T_k)P(t, T_{k+1})L(t, T_k, T_{k+1}) = 0,$$

hence by summation on $k = i, \ldots, j-1$ we get

$$P(t, T_i) - P(t, T_j) - \sum_{k=i}^{j-1} (T_{k+1} - T_k) P(t, T_{k+1}) L(t, T_k, T_{k+1}) = 0.$$

Finally, (10.8) yields

$$
\begin{aligned}
S(t, T_i, T_j) &= \frac{1}{P(t, T_i, T_j)} \sum_{k=i}^{j-1} (T_{k+1} - T_k) P(t, T_{k+1}) L(t, T_k, T_{k+1}) \\
&= \frac{P(t, T_i) - P(t, T_j)}{P(t, T_i, T_j)}.
\end{aligned}
\tag{10.12}
$$

$\square$

Clearly, a simple expression for the swap rate such as that of Proposition 10.4 cannot be obtained using the standard (i.e. non LIBOR) rates defined in (10.11).

The forward swap rate $S(t, T_i, T_j)$ also satisfies

$$P(t, T_i) - P(t, T_j) - S(t, T_i, T_j) \sum_{k=i}^{j-1} (T_{k+1} - T_k) P(t, T_{k+1}) = 0, \quad (10.13)$$

$0 \le t \le T_i,\ 1 \le i < j \le n$.

When $j = i+1$, the swap rate $S(t, T_i, T_{i+1})$ coincides with the forward rate $L(t, T_i, T_{i+1})$:

$$S(t, T_i, T_{i+1}) = L(t, T_i, T_{i+1}), \qquad 1 \le i \le n-1, \tag{10.14}$$

and the bond prices $P(t, T_i)$ can be recovered from the forward swap rates $S(t, T_i, T_j)$ using the relations

$$P(t, T_{i+1}) = \frac{P(t, T_i)}{1 + (T_{i+1} - T_i) S(t, T_i, T_{i+1})},$$

and

$$P(t, T_j) = \frac{P(t, T_i) - S(t, T_i, T_j) \displaystyle\sum_{k=i}^{j-2} (T_k - T_{k-1}) P(t, T_{k+1})}{1 + (T_j - T_{j-1}) S(t, T_i, T_j)},$$

$0 \le t \le T_i,\ 1 \le i < j \le n$.

10.6 Forward Swap Measures

In this section we introduce the forward measures to be used for the pricing of swaptions, and we study their properties.

Definition 10.3. *The forward swap measure* $\mathbb{P}_{i,j}$ *is defined by*

$$\frac{d\mathbb{P}_{i,j}}{d\mathbb{P}} = e^{-\int_0^{T_i} r_s ds} \frac{P(T_i, T_i, T_j)}{P(0, T_i, T_j)},$$

$1 \le i < j \le n.$

We have

$$\mathbb{E}\left[\frac{d\mathbb{P}_{i,j}}{d\mathbb{P}}\Big|\mathcal{F}_t\right] = \frac{1}{P(0, T_i, T_j)} \mathbb{E}\left[e^{-\int_0^{T_i} r_s ds} P(T_i, T_i, T_j)\Big|\mathcal{F}_t\right]$$

$$= \frac{P(t, T_i, T_j)}{P(0, T_i, T_j)} e^{-\int_0^t r_s ds}, \qquad 0 \le t \le T_i,$$

by Proposition 10.3.

Proposition 10.5. *We have*

$$\frac{d\mathbb{P}_{i,j|\mathcal{F}_t}}{d\mathbb{P}_{|\mathcal{F}_t}} = e^{-\int_t^{T_i} r_s ds} \frac{P(T_i, T_i, T_j)}{P(t, T_i, T_j)}, \qquad 0 \le t \le T_{i+1}. \tag{10.15}$$

Proof. It suffices to show that

$$\mathbb{E}_{i,j}[F|\mathcal{F}_t] = \mathbb{E}\left[Fe^{-\int_t^T r_s ds} \frac{P(T_i, T_i, T_j)}{P(t, T_i, T_j)}\Big|\mathcal{F}_t\right] \tag{10.16}$$

for all integrable random variables F.

Now, for any random variable G bounded and $\mathcal{F}_t$-measurable we have

$$\mathbb{E}\left[GFP(T_i, T_i, T_j)e^{-\int_t^T r_s ds}\right] = P(0, T_i, T_j)\,\mathbb{E}_{i,j}\left[GFe^{\int_0^t r_s ds}\right]$$

$$= P(0, T_i, T_j)\,\mathbb{E}_{i,j}\left[Ge^{\int_0^t r_s ds}\,\mathbb{E}_{i,j}[F|\mathcal{F}_t]\right]$$

$$= \mathbb{E}\left[GP(T_i, T_i, T_j)e^{-\int_0^{T_i} r_s ds}e^{\int_0^t r_s ds}\,\mathbb{E}_{i,j}[F|\mathcal{F}_t]\right]$$

$$= \mathbb{E}\left[GP(T_i, T_i, T_j)e^{-\int_t^{T_i} r_s ds}\,\mathbb{E}_{i,j}[F|\mathcal{F}_t]\right]$$

$$= \mathbb{E}\left[G\,\mathbb{E}\left[P(T_i, T_i, T_j)e^{-\int_t^{T_i} r_s ds}\big|\mathcal{F}_t\right]\mathbb{E}_{i,j}[F|\mathcal{F}_t]\right]$$

$$= \mathbb{E}[GP(t, T_i, T_j)\,\mathbb{E}_{i,j}[F|\mathcal{F}_t]],$$

where we used Proposition 10.3, which proves (10.16) by the characterization (12.4) of the conditional expectation. $\qquad\square$

As a consequence of Proposition 10.5 we also have

$$\frac{d\mathbb{P}_{i,j|\mathcal{F}_t}}{d\mathbb{P}_{k|\mathcal{F}_t}} = P(t,T_k)e^{-\int_{T_k}^{T_i} r_s ds}\frac{P(T_i,T_i,T_j)}{P(t,T_i,T_j)}, \qquad 0 \le t \le T_{i+1} \wedge T_k,$$

hence at $t = 0$,

$$\frac{d\mathbb{P}_{i,j}}{d\mathbb{P}_k} = \frac{P(0,T_k)}{P(0,T_i,T_j)}e^{-\int_{T_k}^{T_i} r_s ds}P(T_i,T_i,T_j), \qquad (10.17)$$

$1 \le i < j \le n$. We also have the following result, which extends Propositions 7.2 and 10.2.

Proposition 10.6. *For all $1 \le i < j \le n$ and $1 \le k \le n$ we have*

$$\mathbb{E}_{i,j}\left[\frac{d\mathbb{P}_k}{d\mathbb{P}_{i,j}}\bigg|\mathcal{F}_t\right] = \frac{P(0,T_i,T_j)}{P(0,T_k)}\frac{P(t,T_k)}{P(t,T_i,T_j)} \qquad 0 \le t \le T_{i+1} \wedge T_k, \quad (10.18)$$

and

$$\mathbb{E}_k\left[\frac{d\mathbb{P}_{i,j}}{d\mathbb{P}_k}\bigg|\mathcal{F}_t\right] = \frac{P(0,T_k)}{P(0,T_i,T_j)}\frac{P(t,T_i,T_j)}{P(t,T_k)} \qquad 0 \le t \le T_{i+1} \wedge T_k. \quad (10.19)$$

In particular, the process

$$t \longmapsto v_k^{i,j}(t) := \frac{P(t,T_k)}{P(t,T_i,T_j)}$$

is an $\mathcal{F}_t$-martingale under $\mathbb{P}_{i,j}$, and

$$t \longmapsto v_{i,j}^k(t) := \frac{P(t,T_i,T_j)}{P(t,T_k)}$$

is an $\mathcal{F}_t$-martingale under $\mathbb{P}_k$.

Proof. For all bounded and $\mathcal{F}_t$-measurable random variables F we have

$$\mathbb{E}_{i,j}\left[F\frac{d\mathbb{P}_k}{d\mathbb{P}_{i,j}}\right] = \mathbb{E}\left[F\frac{d\mathbb{P}_k}{d\mathbb{P}}\right]$$

$$= \frac{1}{P(0,T_k)}\mathbb{E}\left[Fe^{-\int_0^{T_k} r_u du}\right]$$

$$= \frac{1}{P(0,T_k)}\mathbb{E}\left[Fe^{-\int_0^t r_u du}P(t,T_k)\right]$$

$$= \frac{1}{P(0,T_k)}\mathbb{E}\left[Fe^{-\int_0^{T_i} r_u du}P(T_i,T_i,T_j)\frac{P(t,T_k)}{P(t,T_i,T_j)}\right]$$

$$= \frac{P(0,T_i,T_j)}{P(0,T_k)}\mathbb{E}_{i,j}\left[F\frac{P(t,T_k)}{P(t,T_i,T_j)}\right],$$

which shows (10.18). Similarly we have

$$\mathbb{E}_k\left[F\frac{d\mathbb{P}_{i,j}}{d\mathbb{P}_k}\right] = \mathbb{E}\left[F\frac{d\mathbb{P}_{i,j}}{d\mathbb{P}}\right]$$

$$= \frac{1}{P(0,T_i,T_j)} \mathbb{E}\left[Fe^{-\int_0^{T_i} r_u du} P(T_i,T_i,T_j)\right]$$

$$= \frac{1}{P(0,T_i,T_j)} \mathbb{E}\left[Fe^{-\int_0^{t} r_u du} P(t,T_i,T_j)\right]$$

$$= \frac{1}{P(0,T_i,T_j)} \mathbb{E}\left[Fe^{-\int_0^{T_k} r_u du} \frac{P(T_k,T_k)}{P(t,T_k)} P(t,T_i,T_j)\right]$$

$$= \frac{P(0,T_k)}{P(0,T_i,T_j)} \mathbb{E}_k\left[F\frac{P(t,T_i,T_j)}{P(t,T_k)}\right],$$

which shows (10.19). The fact that $t \longmapsto v_k^{i,j}(t)$ and $t \longmapsto v_{i,j}^k(t)$ are martingales respectively under $\mathbb{P}_{i,j}$ and $\mathbb{P}_k$ then follows from the remark after Proposition 12.2 in Appendix A. □

It follows from Proposition 10.6 that the swap rate

$$S(t,T_i,T_j) = \frac{P(t,T_i) - P(t,T_j)}{P(t,T_i,T_j)}$$

$$= v_i^{i,j}(t) - v_j^{i,j}(t), \qquad 0 \le t \le T_i,$$

is a martingale under $\mathbb{P}_{i,j}$, using again the tower property (12.7) after Proposition 12.2 in the Appendix A. More precisely, we are able to construct a standard Brownian motion $(B_t^{i,j})_{t\in\mathbb{R}_+}$ under $\mathbb{P}_{i,j}$, which is driving the stochastic evolution of $S(t,T_i,T_j)$, cf. Propositions 10.7 and 10.8 below. Recall that we let $\delta_k = T_{k+1} - T_k$, $k = 1,\ldots,n-1$.

Proposition 10.7. *For all* $i,j = 1,\ldots,n$, *the process*

$$B_t^{i,j} := B_t - \sum_{l=i}^{j-1} \delta_l \int_0^t v_{l+1}^{i,j}(s)\zeta_{l+1}(s)ds, \qquad 0 \le t \le T_{i+1}, \qquad (10.20)$$

is a standard Brownian motion under $\mathbb{P}_{i,j}$.

Proof. By Itô's calculus we have, for any $1 \le i < j \le n$,

$$dv_k^{i,j}(t) = d\left(\frac{P(t,T_k)}{P(t,T_i,T_j)}\right)$$

$$= \frac{dP(t,T_k)}{P(t,T_i,T_j)} - \frac{P(t,T_k)}{P(t,T_i,T_j)^2}dP(t,T_i,T_j) + \frac{P(t,T_k)}{P(t,T_i,T_j)^3}|dP(t,T_i,T_j)|^2$$

$$- \frac{1}{P(t,T_i,T_j)^2}dP(t,T_k) \cdot dP(t,T_i,T_j)$$

$$= \frac{dP(t,T_k)}{P(t,T_i,T_j)} - \frac{P(t,T_k)}{P(t,T_i,T_j)^2}\sum_{l=i}^{j-1}\delta_l dP(t,T_{l+1}) + \frac{P(t,T_k)}{P(t,T_i,T_j)^3}|dP(t,T_i,T_j)|^2$$

$$-\frac{1}{P(t,T_i,T_j)^2}dP(t,T_k)\cdot dP(t,T_i,T_j)$$

$$=\frac{P(t,T_k)}{P(t,T_i,T_j)}(r_t dt + \zeta_k(t)dB_t)$$

$$-\frac{P(t,T_k)}{P(t,T_i,T_j)^2}\sum_{l=i}^{j-1}\delta_l P(t,T_{l+1})(r_t dt + \zeta_{l+1}(t)dB_t)$$

$$+\frac{P(t,T_k)}{P(t,T_i,T_j)^3}\sum_{l,l'=i}^{j-1}\delta_l\delta_{l'}P(t,T_{l+1})P(t,T_{l'+1})\zeta_{l+1}(t)\zeta_{l'+1}(t)dt$$

$$-\frac{P(t,T_k)\zeta_k(t)}{P(t,T_i,T_j)^2}\sum_{l=i}^{j-1}\delta_l\zeta_{l+1}(t)P(t,T_{l+1})dt$$

$$=v_k^{i,j}(t)\left(\zeta_k(t)dB_t - \sum_{l=i}^{j-1}\delta_l v_{l+1}^{i,j}(t)\zeta_{l+1}(t)dB_t\right.$$

$$\left.+\sum_{l,l'=i}^{j-1}\delta_l\delta_{l'}v_{l+1}^{i,j}(t)v_{l'+1}^{i,j}(t)\zeta_{l+1}(t)\zeta_{l'+1}(t)dt - \zeta_k(t)\sum_{l=i}^{j-1}\delta_l\zeta_{l+1}(t)v_{l+1}^{i,j}(t)dt\right)$$

$$=v_k^{i,j}(t)\left(\sum_{l=i}^{j-1}\delta_l v_{l+1}^{i,j}(t)(\zeta_k(t)-\zeta_{l+1}(t))dB_t\right.$$

$$\left.+\sum_{l,l'=i}^{j-1}\delta_l\delta_{l'}v_{l+1}^{i,j}(t)v_{l'+1}^{i,j}(t)(\zeta_{l+1}(t)-\zeta_k(t))\zeta_{l'+1}(t)dt\right)$$

$$=v_k^{i,j}(t)\sum_{l=i}^{j-1}\delta_l v_{l+1}^{i,j}(t)(\zeta_k(t)-\zeta_{l+1}(t))\left(dB_t - \sum_{l'=i}^{j-1}\delta_{l'}v_{l'+1}^{i,j}(t)\zeta_{l'+1}(t)dt\right)$$

$$=v_k^{i,j}(t)\left(\sum_{l=i}^{j-1}\delta_l v_{l+1}^{i,j}(t)(\zeta_k(t)-\zeta_{l+1}(t))\right)dB_t^{i,j}.$$

Since $t \longmapsto v_k^{i,j}(t)$ is a continuous martingale, $B_t^{i,j}$ defined in (10.20) is a standard Brownian motion under $\mathbb{P}_{i,j}$. ☐

When $j = i+1$, Relation (10.20) reads

$$B^{i,i+1}(t) = B_t - \int_0^t \zeta_{i+1}(s)ds, \qquad 0 \le t \le T_{i+1},$$

since $v_{i+1}^{i,i+1}(t) = 1/\delta_i$, hence from Relation (10.4) we have

$$(B_t^{i,i+1})_{t\in[0,T_i]} = (B_t^{i+1})_{t\in[0,T_i]}. \tag{10.21}$$

The equality is valid up to time T_i as by Proposition 10.6 we have

$$\mathbb{E}_{i,i+1}\left[\frac{d\mathbb{P}_{i,i+1}}{d\mathbb{P}_{i+1}}\bigg|\mathcal{F}_t\right] = 1, \qquad 0 \le t \le T_{i+1}, \tag{10.22}$$

although by (10.17) we have

$$\frac{d\mathbb{P}_{i,i+1}}{d\mathbb{P}_{i+1}} = P(T_i, T_{i+1})e^{\int_{T_i}^{T_{i+1}} r_s ds},$$

hence $\mathbb{P}_{i,i+1} \ne \mathbb{P}_{i+1}$.

We can now compute the dynamics of the swap rate $S(t, T_i, T_j)$ under $\mathbb{P}_{i,j}$ using the Brownian process $(B_t^{i,j})_{t \in \mathbb{R}_+}$ as in [Schoenmakers (2005)], page 17.

Proposition 10.8. *We have*

$$dS(t, T_i, T_j) = S(t, T_i, T_j)\sigma_{i,j}(t)dB_t^{i,j}, \qquad 0 \le t \le T_i,$$

where the swap rate volatility is

$$\sigma_{i,j}(t) = \sum_{l=i}^{j-1}\delta_l v_{l+1}^{i,j}(t)(\zeta_i(t) - \zeta_{l+1}(t)) + \frac{P(t, T_j)}{P(t, T_i) - P(t, T_j)}(\zeta_i(t) - \zeta_j(t)),$$

$1 \le i, j \le n.$

Proof. From the proof of Proposition 10.7 we have

$$dv_k^{i,j}(t) = v_k^{i,j}(t)\left(\sum_{l=i}^{j-1}\delta_l v_{l+1}^{i,j}(t)(\zeta_k(t) - \zeta_{l+1}(t))\right)dB_t^{i,j},$$

hence

$$dS(t, T_i, T_j) = d\left(\frac{P(t, T_i) - P(t, T_j)}{P(t, T_i, T_j)}\right)$$

$$= dv_i^{i,j}(t) - dv_j^{i,j}(t)$$

$$= \left(\sum_{l=i}^{j-1}\delta_l v_{l+1}^{i,j}(t)(v_i^{i,j}(t)(\zeta_i(t) - \zeta_{l+1}(t)) - v_j^{i,j}(t)(\zeta_j(t) - \zeta_{l+1}(t)))\right)dB_t^{i,j}$$

$$= \left(\sum_{l=i}^{j-1}\zeta_{l+1}(t)\delta_l v_{l+1}^{i,j}(t)(v_j^{i,j}(t) - v_i^{i,j}(t))\right)dB_t^{i,j}$$

$$+ \left(v_i^{i,j}(t)\zeta_i(t) - v_j^{i,j}(t)\zeta_j(t)\right)dB_t^{i,j}$$

$$= \left(\sum_{l=i}^{j-1}(\zeta_{l+1}(t) - \zeta_i(t))\delta_l v_{l+1}^{i,j}(t)(v_j^{i,j}(t) - v_i^{i,j}(t))\right)dB_t^{i,j}$$

$$+v_j^{i,j}(t)\left(\zeta_i(t)-\zeta_j(t)\right)dB_t^{i,j}$$

$$= S(t,T_i,T_j)\left(\sum_{l=i}^{j-1}\delta_l v_{l+1}^{i,j}(t)(\zeta_i(t)-\zeta_{l+1}(t))\right.$$

$$\left.+\frac{P(t,T_j)}{P(t,T_i)-P(t,T_j)}(\zeta_i(t)-\zeta_j(t))\right)dB_t^{i,j}$$

$$= S(t,T_i,T_j)\sigma_{i,j}(t)dB_t^{i,j}. \qquad\qquad\qquad \square$$

As a consequence of Proposition 10.8 and Corollary 1.1 we recover the fact that the swap rate $S(t,T_i,T_j)$ is a martingale under the forward swap measure $\mathbb{P}_{i,j}$.

The swap rate dynamics derived in Proposition 10.8 will be used in Chapter 11 to compute swaption prices using the Black-Scholes formula under a suitable approximation of the swap rate volatility $\sigma_{i,j}(t)$.

10.7   Swaption Pricing on the LIBOR Market

The relation

$$\sum_{k=i}^{j-1}(T_{k+1}-T_k)P(t,T_{k+1})(L(t,T_k,T_{k+1})-S(t,T_i,T_j))$$

$$= P(t,T_i)-P(t,T_j)-S(t,T_i,T_j)\sum_{k=i}^{j-1}(T_{k+1}-T_k)P(t,T_{k+1})$$

$$= 0,$$

defining the forward swap rate $S(t,T_i,T_j)$ shows that

$$\sum_{k=i}^{j-1}(T_{k+1}-T_k)P(t,T_{k+1})L(t,T_k,T_{k+1})$$

$$= S(t,T_i,T_j)\sum_{k=i}^{j-1}(T_{k+1}-T_k)P(t,T_{k+1})$$

$$= P(t,T_i,T_j)S(t,T_i,T_j)$$

$$= P(t,T_i)-P(t,T_j),$$

by the definition (10.6) of $P(t,T_i,T_j)$, hence

$$\sum_{k=i}^{j-1}(T_{k+1}-T_k)P(t,T_{k+1})(L(t,T_k,T_{k+1})-\kappa)$$

$$= P(t, T_i, T_j)\left(S(t, T_i, T_j) - \kappa\right).$$

In particular for $t = T_i$ we get

$$\left(\sum_{k=i}^{j-1}(T_{k+1} - T_k)P(T_i, T_{k+1})(L(T_i, T_k, T_{k+1}) - \kappa)\right)^+$$

$$= (P(T_i, T_i) - P(T_i, T_j) - \kappa P(T_i, T_i, T_j))^+$$

$$= P(T_i, T_i, T_j)\left(S(T_i, T_i, T_j) - \kappa\right)^+.$$

As a consequence, a swaption on the LIBOR market can be priced from (10.10) at time $t \in [0, T_i]$ using the forward measure $\mathbb{P}_i$ as

$$\mathbb{E}\left[e^{-\int_t^{T_i} r_s ds}\left(\sum_{k=i}^{j-1}(T_{k+1} - T_k)P(T_i, T_{k+1})(L(T_i, T_k, T_{k+1}) - \kappa)\right)^+\bigg|\mathcal{F}_t\right]$$

$$= \mathbb{E}\left[e^{-\int_t^{T_i} r_s ds}\left(P(T_i, T_i) - P(T_i, T_j) - \kappa P(T_i, T_i, T_j)\right)^+\bigg|\mathcal{F}_t\right]$$

$$= \mathbb{E}\left[e^{-\int_t^{T_i} r_s ds}P(T_i, T_i, T_j)\left(S(T_i, T_i, T_j) - \kappa\right)^+\bigg|\mathcal{F}_t\right]$$

$$= P(t, T_i)\mathbb{E}_i\left[P(T_i, T_i, T_j)\left(S(T_i, T_i, T_j) - \kappa\right)^+\bigg|\mathcal{F}_t\right].$$

Finally, by Proposition 10.5 we can also write

$$\mathbb{E}\left[e^{-\int_t^{T_i} r_s ds}\left(\sum_{k=i}^{j-1}(T_{k+1} - T_k)P(T_i, T_{k+1})(L(T_i, T_k, T_{k+1}) - \kappa)\right)^+\bigg|\mathcal{F}_t\right]$$

$$= \mathbb{E}\left[e^{-\int_t^{T_i} r_s ds}P(T_i, T_i, T_j)\left(S(T_i, T_i, T_j) - \kappa\right)^+\bigg|\mathcal{F}_t\right]$$

$$= P(t, T_i, T_j)\mathbb{E}_{i,j}\left[\left(S(T_i, T_i, T_j) - \kappa\right)^+\bigg|\mathcal{F}_t\right], \tag{10.23}$$

where the forward measure $\mathbb{P}_{i,j}$ is defined in (10.15).

10.8 Exercises

Exercise 10.1. Consider a market with three zero-coupon bonds with prices $P(t, T_1)$, $P(t, T_2)$ and $P(t, T_3)$ with maturities $T_1 = \delta$, $T_2 = 2\delta$ and $T_3 = 3\delta$ respectively, and the forward LIBOR $L(t, T_1, T_2)$ and $L(t, T_2, T_3)$ defined by

$$L(t, T_i, T_{i+1}) = \frac{1}{\delta}\left(\frac{P(t, T_i)}{P(t, T_{i+1})} - 1\right), \qquad i = 1, 2.$$

Assume that $L(t, T_1, T_2)$ and $L(t, T_2, T_3)$ are modeled as

$$\frac{dL(t, T_1, T_2)}{L(t, T_1, T_2)} = \gamma_1(t)dB_t^{(2)}, \qquad 0 \le t \le T_1, \qquad (10.24)$$

and $L(t, T_2, T_3) = b$, $0 \le t \le T_2$, for some constant $b > 0$ and function $\gamma_1(t)$, where $B_t^{(2)}$ is a standard Brownian motion under the forward measure $\mathbb{P}_2$ defined by

$$\frac{d\mathbb{P}_2}{d\mathbb{P}} = \frac{e^{-\int_0^{T_2} r_s ds}}{P(0, T_2)}.$$

(1) Compute $L(t, T_1, T_2)$, $0 \le t \le T_2$ by solving Equation (10.24).
(2) Compute the prices at time t:

$$\mathbb{E}\left[e^{-\int_t^{T_{i+1}} r_s ds} (L(T_i, T_i, T_{i+1}) - \kappa)^+ \Big| \mathcal{F}_t\right]$$
$$= P(t, T_i)\, \mathbb{E}_{i+1}\left[(L(T_i, T_i, T_{i+1}) - \kappa)^+ \mid \mathcal{F}_t\right], \qquad 0 \le t \le T_i,$$

of the caplets with strike κ, where $\mathbb{E}_{i+1}$ denotes the expectation under the forward measure $\mathbb{P}_{i+1}$, $i = 1, 2$.
(3) Compute

$$\frac{P(t, T_1)}{P(t, T_1, T_3)}, \quad 0 \le t \le T_1, \quad \text{and} \quad \frac{P(t, T_3)}{P(t, T_1, T_3)}, \quad 0 \le t \le T_2,$$

in terms of b and $L(t, T_1, T_2)$, where $P(t, T_1, T_3)$ is the annuity numeraire

$$P(t, T_1, T_3) = \delta P(t, T_2) + \delta P(t, T_3), \qquad 0 \le t \le T_2.$$

(4) Compute the dynamics of the swap rate

$$t \longmapsto S(t, T_1, T_3) = \frac{P(t, T_1) - P(t, T_3)}{P(t, T_1, T_3)}, \qquad 0 \le t \le T_1,$$

i.e. show that we have

$$dS(t, T_1, T_3) = \sigma_{1,3}(t)S(t, T_1, T_3)dB_t^{(2)},$$

where $\sigma_{1,3}(t)$ is a process to be determined.

Exercise 10.2. Consider a market with short term interest rate $(r_t)_{t \in \mathbb{R}_+}$ and two zero-coupon bonds $P(t, T_1)$, $P(t, T_2)$ with maturities $T_1 = \delta$ and $T_2 = 2\delta$, where $P(t, T_i)$ is modeled according to

$$\frac{dP(t, T_i)}{P(t, T_i)} = r_t dt + \zeta_i(t)dB_t, \qquad i = 1, 2.$$

Consider also the forward LIBOR $L(t, T_1, T_2)$ defined by

$$L(t, T_1, T_2) = \frac{1}{\delta} \left(\frac{P(t, T_1)}{P(t, T_2)} - 1 \right), \qquad 0 \le t \le T_1,$$

and assume that $L(t, T_1, T_2)$ is modeled in the BGM model as

$$\frac{dL(t, T_1, T_2)}{L(t, T_1, T_2)} = \gamma dB_t^{(2)}, \qquad 0 \le t \le T_1, \tag{10.25}$$

where γ is a deterministic constant, and

$$B_t^{(2)} = B_t - \int_0^t \zeta_2(s) ds$$

is a standard Brownian motion under the forward measure $\mathbb{P}_2$ defined by

$$\frac{d\mathbb{P}_2}{d\mathbb{P}} = \exp\left(\int_0^{T_2} \zeta_2(s) dB_s - \frac{1}{2} \int_0^{T_2} |\zeta_2(s)|^2 ds \right).$$

(1) Compute $L(t, T_1, T_2)$ by solving Equation (10.25).
(2) Compute the price at time t:

$$P(t, T_2) \mathbb{E}_2 \left[(L(T_1, T_1, T_2) - \kappa)^+ \mid \mathcal{F}_t \right], \qquad 0 \le t \le T_1,$$

of the caplet with strike κ, where $\mathbb{E}_2$ denotes the expectation under the forward measure $\mathbb{P}_2$.

Exercise 10.3. (Exercise 7.3 continued).

(1) Compute the dynamics of the swap rate process

$$S(t, T_1, T_2) = \frac{P(t, T_1) - P(t, T_2)}{(T_2 - T_1) P(t, T_2)}, \qquad t \in [0, T_1],$$

under $\mathbb{P}_2$.
(2) Compute the swaption price

$$(T_2 - T_1) \mathbb{E} \left[e^{-\int_t^{T_1} r_s ds} P(T_1, T_2)(L(T_1, T_1, T_2) - \kappa)^+ \Big| \mathcal{F}_t \right]$$

on the LIBOR rate $L(T_1, T_1, T_2)$ using the expectation under the forward swap measure $\mathbb{P}_{1,2}$.

Exercise 10.4. Swaption hedging. Consider a bond market with tenor structure $\{T_i, \ldots, T_j\}$ and bonds with maturities $T_i, \ldots, T_j$, whose prices $P(t, T_i), \ldots, P(t, T_j)$ at time t are given by

$$\frac{dP(t, T_k)}{P(t, T_k)} = r_t dt + \zeta_k(t) dB_t, \qquad k = i, \ldots, j,$$

where $(r_t)_{t \in \mathbb{R}_+}$ is a short term interest rate process, $(B_t)_{t \in \mathbb{R}_+}$ is a standard Brownian motion generating the filtration $(\mathcal{F}_t)_{t \in \mathbb{R}_+}$, and $\zeta_i(t), \ldots, \zeta_j(t)$ are volatility processes. The swap rate $S(t, T_i, T_j)$ is defined as in (10.12) by $S(t, T_i, T_j) = (P(t, T_i) - P(t, T_j))/P(t, T_i, T_j)$, where $P(t, T_i, T_j)$ is the annuity numeraire (10.6).

(1) Assume that the swap rate is modeled as a geometric Brownian motion

$$dS(t, T_i, T_j) = S(t, T_i, T_j)\sigma_{i,j}(t)dB_t^{i,j}, \qquad 0 \leq t \leq T_i, \qquad (10.26)$$

where the swap rate volatility is a deterministic function $\sigma_{i,j}(t)$ and $B_t^{i,j}$ is a standard Brownian motion under the swap measure $\mathbb{P}_{i,j}$, cf. Proposition 10.7. Show that the price at time $t \in [0, T_i]$ of a swaption on the LIBOR market can be written as

$$\mathbb{E}\left[e^{-\int_t^{T_i} r_s ds} \left(\sum_{k=i}^{j-1} (T_{k+1} - T_k) P(T_i, T_{k+1})(L(T_i, T_k, T_{k+1}) - \kappa) \right)^+ \Big| \mathcal{F}_t \right]$$

$$= P(t, T_i, T_j) \, \mathbb{E}_{i,j}\left[(S(T_i, T_i, T_j) - \kappa)^+ \Big| \mathcal{F}_t \right]$$

$$= P(t, T_i, T_j) C(S(t, T_i, T_j), \kappa, v(t, T_i)),$$

under the forward swap measure $\mathbb{P}_{i,j}$ of Definition 10.3,

$$v^2(t, T_i) = \int_t^{T_i} |\sigma_{i,j}(s)|^2 ds,$$

and $C(x, \kappa, v)$ is a function to be specified using the Black-Scholes formula of Lemma 2.3.

(2) Consider a portfolio $(\xi_t^i, \ldots, \xi_t^j)_{t \in [0, T_i]}$ made of bonds with maturities $T_i, \ldots, T_j$, and value

$$V_t = \sum_{k=i}^{j} \xi_t^k P(t, T_k),$$

at time $t \in [0, T_i]$. We assume that the portfolio is self-financing, i.e.

$$dV_t = \sum_{k=i}^{j} \xi_t^k dP(t, T_k), \qquad 0 \leq t \leq T_i, \qquad (10.27)$$

and that it *hedges* the claim $(S(T_i, T_i, T_j) - \kappa)^+$, so that

$$V_t = P(t, T_i, T_j) \, \mathbb{E}_{i,j}\left[(S(T_i, T_i, T_j) - \kappa)^+ \Big| \mathcal{F}_t \right], \qquad 0 \leq t \leq T_i.$$

Show that the forward portfolio price $\hat{V}_t = V_t/P(t, T_i, T_j)$ satisfies

$$d\hat{V}_t = \frac{\partial C}{\partial x}(S_t, \kappa, v(t, T_i))dS_t.$$

(3) Show that we have

$$dV_t = \frac{\partial C}{\partial x}(S_t, \kappa, v(t, T_i))d(P(t, T_i) - P(t, T_j))$$
$$+ \left(\hat{V}_t - S_t \frac{\partial C}{\partial x}(S_t, \kappa, v(t, T_i))\right) dP(t, T_i, T_j).$$

(4) Compute the hedging strategy $(\xi_t^i, \ldots, \xi_t^j)$ of the swaption.

Chapter 11

The Brace-Gatarek-Musiela (BGM) Model

This chapter is devoted to the BGM model, a nonlinear model for LIBOR rates which, unlike the HJM model, ensures the positivity of interest rates, which can be built according to a drifted geometric Brownian motion. After constructing the BGM model we give a quick outlook on its calibration following the approach of [Schoenmakers (2005)].

11.1 The BGM Model

The BGM model has been introduced in [Brace *et al.* (1997)] for the pricing of interest rate derivatives such as caps and swaptions on the LIBOR market.

The models (HJM, affine, etc.) considered in the previous chapter suffer from the following drawbacks:

- explicitly computable models such as the Vasicek model do not satisfy the positivity of rates property.

- models with positive rates (e.g. the CIR model) do not lead to explicit analytical formulas.

- the lack of explicit analytical formulas makes it necessary to use the Monte Carlo method for pricing, which makes model calibration difficult in practice.

- fitting the forward interest rate curves in these models is problematic.

Thus there is a strong interest in searching for models that:

- yield positive interest rates, and

- permit to derive explicit formulas for the computation of prices.

These two goals can be achieved by the BGM model. Assume that the bond price $P(t, T_i)$ satisfies

$$\frac{dP(t, T_i)}{P(t, T_i)} = r_t dt + \zeta_i(t) dB_t, \qquad i = 1, \ldots, n, \qquad (11.1)$$

where B_t is a standard Brownian motion under $\mathbb{P}$.

For $i = 1, \ldots, n$, the process

$$B_t^i := B_t - \int_0^t \zeta_i(s) ds, \qquad 0 \leq t \leq T_i,$$

is a standard $\mathbb{R}^d$-valued Brownian motion under the probability $\mathbb{P}_i$ defined in Definition 10.1.

In the BGM model we assume that $L(t, T_i, T_{i+1})$ is a geometric Brownian motion under $\mathbb{P}_{i+1}$, i.e.

$$\frac{dL(t, T_i, T_{i+1})}{L(t, T_i, T_{i+1})} = \gamma_i(t) dB_t^{i+1}, \qquad (11.2)$$

$0 \leq t \leq T_i$, $i = 1, \ldots, n-1$, for some deterministic function $\gamma_i(t)$, $i = 1, \ldots, n-1$, with solution

$$L(u, T_i, T_{i+1}) = L(t, T_i, T_{i+1}) \exp\left(\int_t^u \gamma_i(s) dB_s^{i+1} - \frac{1}{2} \int_t^u |\gamma_i|^2(s) ds\right),$$

i.e. for $u = T_i$,

$$L(T_i, T_i, T_{i+1}) = L(t, T_i, T_{i+1}) \exp\left(\int_t^{T_i} \gamma_i(s) dB_s^{i+1} - \frac{1}{2} \int_t^{T_i} |\gamma_i|^2(s) ds\right).$$

Since $L(t, T_i, T_{i+1})$ is a geometric Brownian motion under $\mathbb{P}_{i+1}$, $i = 0, \ldots, n-1$, standard caplets can be priced at time $t \in [0, T_i]$ from the Black-Scholes formula of Section 2.3, see Section 11.2 below.

Let us now determine the dynamics of $L(t, T_i, T_{i+1})$ under $\mathbb{P}$. Again we let

$$\delta_k = T_{k+1} - T_k, \qquad k = 1, \ldots, n-1.$$

Proposition 11.1. *For $1 \leq i < n$ we have*

$$\frac{dL(t, T_i, T_{i+1})}{L(t, T_i, T_{i+1})} = -\gamma_i(t) \sum_{j=i+1}^{n-1} \frac{\delta_j \gamma_j(t) L(t, T_j, T_{j+1})}{1 + \delta_j L(t, T_j, T_{j+1})} dt + \gamma_i(t) dB_t^n, \quad (11.3)$$

$0 \leq t \leq T_i$, *where $\gamma_i(t)$ is a deterministic function, $(B_t^n)_{t \in \mathbb{R}_+}$ is a standard Brownian motion under $\mathbb{P}_n$, and $L(t, T_i, T_{i+1})$, $0 \leq t \leq T_i$, is a martingale under $\mathbb{P}_{i+1}$, $i = 1, \ldots, n-1$.*

Proof. We have

$$d\left(\frac{P(t, T_i)}{P(t, T_{i+1})}\right) = d(1 + \delta_i L(t, T_i, T_{i+1}))$$

$$= \delta_i L(t, T_i, T_{i+1}) \gamma_i(t) dB_t^{i+1}$$

$$= \frac{P(t, T_i)}{P(t, T_{i+1})} \frac{\delta_i L(t, T_i, T_{i+1})}{1 + \delta_i L(t, T_i, T_{i+1})} \gamma_i(t) dB_t^{i+1}. \quad (11.4)$$

On the other hand, using the dynamics

$$\frac{dP(t, T_i)}{P(t, T_i)} = r_t dt + \zeta_i(t) dB_t, \qquad i = 1, \ldots, n,$$

and Itô's calculus, we have

$$d\left(\frac{P(t, T_i)}{P(t, T_{i+1})}\right) = \frac{P(t, T_i)}{P(t, T_{i+1})}(\zeta_i(t) - \zeta_{i+1}(t))(dB_t - \zeta_{i+1}(t)dt)$$

$$= \frac{P(t, T_i)}{P(t, T_{i+1})}(\zeta_i(t) - \zeta_{i+1}(t))dB_t^{i+1}. \quad (11.5)$$

By identification of (11.4) with (11.5) we get

$$\zeta_{i+1}(t) - \zeta_i(t) = -\frac{\delta_i L(t, T_i, T_{i+1}) \gamma_i(t)}{1 + \delta_i L(t, T_i, T_{i+1})}, \quad (11.6)$$

$0 \leq t \leq T_i$, $i = 1, \ldots, n-1$, and

$$\frac{dL(t, T_i, T_{i+1})}{L(t, T_i, T_{i+1})} = \gamma_i(t)dB_t^{i+1}$$

$$= \gamma_i(t)dB_t^i + \frac{\delta_i L(t, T_i, T_{i+1})|\gamma_i|^2(t)}{1 + \delta_i L(t, T_i, T_{i+1})}dt,$$

since $dB_t^{i+1} = dB_t - \zeta_{i+1}(t)dt$ and

$$dB_t^{i+1} = dB_t^i - (\zeta_{i+1}(t) - \zeta_i(t))dt, \qquad 1 \leq i \leq n-1.$$

Hence from (11.6) we have

$$\zeta_k(t) = \zeta_i(t) - \sum_{j=i}^{k-1} \frac{\delta_j L(t, T_j, T_{j+1}) \gamma_j(t)}{1 + \delta_j L(t, T_j, T_{j+1})}, \quad (11.7)$$

$0 \leq t \leq T_i$, $1 \leq i < k \leq n$, and for $k > i$,

$$\frac{dL(t, T_i, T_{i+1})}{L(t, T_i, T_{i+1})} = \gamma_i(t)dB_t^{i+1}$$

$$= \gamma_i(t)dB_t^k - \gamma_i(t)(dB_t^k - dB_t^{i+1})$$

$$= \gamma_i(t)dB_t^k - \gamma_i(t) \sum_{j=i+1}^{k-1} (\zeta_j(t) - \zeta_{j+1}(t))dt$$

$$= - \sum_{j=i+1}^{k-1} \frac{\delta_j L(t, T_j, T_{j+1})}{1 + \delta_j L(t, T_j, T_{j+1})} \gamma_i(t)\gamma_j(t)dt + \gamma_i(t)dB_t^k,$$

$0 \leq t \leq T_i$. $\qquad\qquad\qquad\qquad\qquad\qquad\qquad\qquad\qquad\qquad\qquad\quad\square$

Similarly, for $1 \leq k \leq i < n$ we have:

$$\frac{dL(t, T_i, T_{i+1})}{L(t, T_i, T_{i+1})} = \gamma_i(t)dB_t^k + \gamma_i(t) \sum_{j=k}^{i}(\zeta_j(t) - \zeta_{j+1}(t))dt$$

$$= \gamma_i(t)dB_t^k + \gamma_i(t) \sum_{j=k}^{i} \frac{\delta_j \gamma_j(t)L(t, T_j, T_{j+1})}{1 + \delta_j L(t, T_j, T_{j+1})}dt.$$

11.2   Cap Pricing

As a consequence of Relations (10.1) and (11.2) and using the Black-Scholes formula, the caplet of payoff

$$(L(T_i, T_i, T_{i+1}) - \kappa)^+$$

can be priced as time $t \in [0, T_i]$ as

$$\mathbb{E}\left[e^{-\int_t^{T_{i+1}} r_s ds}(L(T_i, T_i, T_{i+1}) - \kappa)^+ \Big| \mathcal{F}_t\right]$$

$$= P(t, T_{i+1})\, \mathbb{E}_{i+1}\left[(L(T_i, T_i, T_{i+1}) - \kappa)^+ \mid \mathcal{F}_t\right]$$

$$= P(t, T_{i+1})\mathrm{Bl}(\kappa, L(t, T_i, T_{i+1}), \sigma_i(t), 0, T_i - t),$$

where $\mathrm{Bl}(\kappa, x, \sigma, r, \tau)$ is the Black-Scholes function defined in Section 2.3, with

$$|\sigma_i(t)|^2 = \frac{1}{T_i - t} \int_t^{T_i} |\gamma_i|^2(s)ds. \qquad (11.8)$$

Vol Cap At the Money	1M	3M	6M	12M	2Y	3Y	4Y	5Y	7Y	10Y
2D	9,25	9	8,85	18,6	18	16,8	15,7	14,7	13	11,3
1M	15,35	15,1	14,95	17,6	18,03	16,83	15,73	14,73	13,03	11,33
2M	15,75	15,5	15,35	18,1	18,41	17,11	16,01	15,01	13,26	11,56
3M	15,55	15,3	15,15	18,6	18,79	17,39	16,29	15,29	13,49	11,79
6M	17,55	17,3	17,15	18,7	18,28	16,98	15,88	14,98	13,48	11,98
9M	18,35	18,1	17,95	18,3	17,76	16,56	15,51	14,66	13,31	12,01
1Y	19,25	19	18,85	17,9	17,25	16,15	15,15	15,15	13,15	12,05
2Y	17,85	17,6	17,45	16,3	15,96	15,16	14,46	13,86	12,96	12,06
3Y	16,8	16,55	16,4	15,2	15,38	14,58	13,98	13,58	12,88	12,18
4Y	15,6	15,35	15,2	14,4	14,79	14,19	13,69	13,29	12,79	12,29
5Y	14,65	14,4	14,25	13,4	14,5	13,97	13,53	13,2	12,8	12,4
6Y	13,8	13,55	13,45	12,85	14,19	13,66	13,17	12,89	12,54	12,14
7Y	13,35	13,1	13	12,3	13,88	13,35	12,81	12,58	12,28	11,88
8Y	13,1	12,85	12,75	11,97	13,65	13,15	12,65	12,42	12,12	11,75
9Y	12,75	12,5	12,4	11,63	13,43	12,96	12,49	12,26	11,96	11,63
10Y	12,4	12,15	12,05	11,3	13,5	13,02	12,53	12,25	11,89	11,5
12Y	11,85	11,6	11,5	10,8	13,22	12,75	12,28	12,01	11,69	11,3
15Y	11,25	11	10,9	10,2	13	12,55	12,1	11,85	11,57	11,15
20Y	10,45	10,2	10,1	9,5	11,9	11,55	11,2	11,05	11,03	10,8
25Y	9,7	9,45	9,35	8,8	11,68	11,33	10,98	10,83	10,88	10,55
30Y	9,05	8,8	8,7	8,1	11,45	11,1	10,75	10,6	10,72	10,3

Fig. 11.1   Caplet volatilities.

By inversion of the Black-Scholes formula one can compute implied caplet volatilities $\sigma_i^B(t)$ from market data. The table[1] given in Figure 11.1 presents such implied volatilities, where the time to maturity $T_i - t$ is in ordinate and the period $T_j - T_i$ is in abscissa.

The pricing of caplets extends to caps with payoff

$$\sum_{k=i}^{j-1} (T_{k+1} - T_k)(L(T_k, T_k, T_{k+1}) - \kappa)^+$$

since they can be decomposed into a sum of caplets, and priced at time $t \in [0, T_i]$ as

$$\sum_{k=i}^{j-1} \delta_k P(t, T_{k+1}) \mathrm{Bl}(\kappa, L(t, T_k, T_{k+1}), \sigma_k(t), 0, T_k - t).$$

11.3   Swaption Pricing

We already know that the swaption with payoff

$$\left(\sum_{k=i}^{j-1} \delta_k P(T_i, T_{k+1})(L(T_i, T_k, T_{k+1}) - \kappa) \right)^+$$

---

[1]Data communicated to the author by Natixis Corporate & Investment Bank.

on the LIBOR market is priced at time $t \in [0, T_i]$ as

$$P(t, T_i) \, \mathbb{E}_i \left[P(T_i, T_i, T_j) \left(S(T_i, T_i, T_j) - \kappa \right)^+ \Big| \mathcal{F}_t \right]$$

$$= P(t, T_i, T_j) \, \mathbb{E}_{i,j} \left[\left(S(T_i, T_i, T_j) - \kappa \right)^+ \Big| \mathcal{F}_t \right], \qquad (11.9)$$

where the martingale measure $\mathbb{P}_{i,j}$ has been defined in (10.15) by

$$\frac{d\mathbb{P}_{i,j|\mathcal{F}_t}}{d\mathbb{P}_{|\mathcal{F}_t}} = e^{-\int_t^{T_i} r_s ds} \frac{P(T_i, T_i, T_j)}{P(t, T_i, T_j)}, \qquad 0 \le t \le T_i,$$

$1 \le i < j \le n$, cf. Section 10.7.

Swaption prices can be computed by the Monte Carlo method using the dynamics of $L(t, T_k, T_{k+1})$ under $\mathbb{P}_i$, $1 \le i \le k \le n$, but the market practice is to use approximation formulas. Recall that the swap rate $S(t, T_i, T_j)$ satisfies

$$S(t, T_i, T_j) = \frac{1}{P(t, T_i, T_j)} \sum_{k=i}^{j-1} (T_{k+1} - T_k) P(t, T_{k+1}) L(t, T_k, T_{k+1}), \quad (11.10)$$

where

$$P(t, T_i, T_j) = \sum_{k=i}^{j-1} (T_{k+1} - T_k) P(t, T_{k+1})$$

is the annuity numeraire. Moreover, the process $v_k^{i,j}$ defined by

$$t \longmapsto v_k^{i,j}(t) := \frac{P(t, T_k)}{P(t, T_i, T_j)}, \qquad 0 \le t \le T_i \wedge T_j,$$

is an $\mathcal{F}_t$-martingale under $\mathbb{P}_{i,j}$, $1 \le i, j \le n$, and by Propositions 10.6 and 10.7 we have

$$dv_k^{i,j}(t) = v_k^{i,j}(t) \left(\sum_{l=i}^{j-1} \delta_l v_{l+1}^{i,j}(t) (\zeta_k(t) - \zeta_{l+1}(t)) \right) dB_t^{i,j},$$

where for all $i, j = 1, \ldots, n$, the process

$$B_t^{i,j} := B_t - \sum_{k=i}^{j-1} \delta_k \int_0^t v_{k+1}^{i,j}(s) \zeta_{k+1}(s) ds, \qquad 0 \le t \le T_i,$$

is a standard Brownian motion under $\mathbb{P}_{i,j}$.

Recall also that by Proposition 10.8 we have

$$dS(t, T_i, T_j) = S(t, T_i, T_j) \sigma_{i,j}(t) dB_t^{i,j},$$

where, using Relation (11.7), the swap rate volatility $\sigma_{i,j}(t)$ can be computed as

$$\sigma_{i,j}(t) = \sum_{l=i}^{j-1} \left(\delta_l v_{l+1}^{i,j}(t)(\zeta_i(t) - \zeta_{l+1}(t)) + \frac{P(t,T_j)}{P(t,T_i) - P(t,T_j)}(\zeta_i(t) - \zeta_j(t)) \right)$$

$$= \sum_{l=i}^{j-1} \delta_l v_{l+1}^{i,j}(t) \sum_{k=i}^{l} \frac{\gamma_k(t)\delta_k L(t,T_k,T_{k+1})}{1 + \delta_k L(t,T_k,T_{k+1})}$$

$$+ \frac{P(t,T_j)}{P(t,T_i) - P(t,T_j)} \sum_{k=i}^{j-1} \frac{\gamma_k(t)\delta_k L(t,T_k,T_{k+1})}{1 + \delta_k L(t,T_k,T_{k+1})}$$

$$= \sum_{k=i}^{j-1} \frac{\gamma_k(t)\delta_k L(t,T_k,T_{k+1})}{1 + \delta_k L(t,T_k,T_{k+1})} \sum_{l=k}^{j-1} \left(\delta_l v_{l+1}^{i,j}(t) + \frac{P(t,T_j)}{P(t,T_i) - P(t,T_j)} \right)$$

$$= \frac{1}{S(t,T_i,T_j)} \times$$

$$\sum_{k=i}^{j-1} \frac{\gamma_k(t)\delta_k L(t,T_k,T_{k+1})}{1 + \delta_k L(t,T_k,T_{k+1})} \sum_{l=k}^{j-1} \left(\delta_l v_{l+1}^{i,j}(t) \frac{P(t,T_i) - P(t,T_j)}{P(t,T_i,T_j)} + \frac{P(t,T_j)}{P(t,T_i,T_j)} \right)$$

$$= \frac{1}{S(t,T_i,T_j)} \sum_{k=i}^{j-1} \gamma_k(t) w_k^{i,j}(t) L(t,T_k,T_{k+1}),$$

with

$$w_k^{i,j}(t) = \frac{\delta_k}{1 + \delta_k L(t,T_k,T_{k+1})} \left(\sum_{l=k}^{j-1} \delta_l v_{l+1}^{i,j}(t) \frac{P(t,T_i) - P(t,T_j)}{P(t,T_i,T_j)} + \frac{P(t,T_j)}{P(t,T_i,T_j)} \right),$$

and $w_i^{i,i+1}(t) = 1$, hence

$$\sigma_{i,j}^2(t) = \frac{1}{S(t,T_i,T_j)^2} \sum_{l=i}^{j-1}\sum_{k=i}^{j-1} \gamma_l(t)\gamma_k(t) w_l^{i,j}(t) w_k^{i,j}(t) L(t,T_l,T_{l+1}) L(t,T_k,T_{k+1}),$$

(11.11)

and $\sigma_{i,i+1}(t) = \gamma_i(t)$, $i = 1, \ldots, n-1$. When $j = i+1$, although

$$\frac{d\mathbb{P}_{i,i+1|\mathcal{F}_t}}{d\mathbb{P}_{|\mathcal{F}_t}} = e^{-\int_t^{T_i} r_s ds} \frac{P(T_i,T_i,T_{i+1})}{P(t,T_i,T_{i+1})}$$

$$= e^{-\int_t^{T_i} r_s ds} \frac{P(T_i,T_{i+1})}{P(t,T_{i+1})}$$

$$\neq \frac{e^{-\int_t^{T_{i+1}} r_s ds}}{P(t,T_{i+1})}$$

$$= \frac{d\mathbb{P}_{i+1|\mathcal{F}_t}}{d\mathbb{P}_{|\mathcal{F}_t}}, \qquad 0 \le t \le T_i,$$

we know by (10.21), (10.22) that $(B_t^{i,i+1})_{t\in[0,T_i]}$ and $(B_t^{i+1})_{t\in[0,T_i]}$ coincide up to time T_i, and by (10.14) the swaption price (11.9) can be computed as

$$P(t, T_i, T_{i+1})\, \mathbb{E}_{i,i+1}\left[\left(S(T_i, T_i, T_{i+1}) - \kappa\right)^+ \Big| \mathcal{F}_t\right]$$

$$= (T_{i+1} - T_i) P(t, T_{i+1})\, \mathbb{E}_{i+1}\left[\left(S(T_i, T_i, T_{i+1}) - \kappa\right)^+ \Big| \mathcal{F}_t\right]$$

$$= (T_{i+1} - T_i) P(t, T_{i+1})\, \mathbb{E}_{i+1}\left[\left(L(T_i, T_i, T_{i+1}) - \kappa\right)^+ \Big| \mathcal{F}_t\right],$$

which is equal to

$$(T_{i+1} - T_i) P(t, T_{i+1}) \mathrm{Bl}(\kappa, L(t, T_i, T_{i+1}), \sigma_i(t), 0, T_i - t), \qquad (11.12)$$

where $\sigma_i(t)$ is defined in (11.8). The next proposition extends this relation to general indices $1 \leq i < j \leq n$ as an approximation, known as the swaption approximation formula.

Proposition 11.2. *The swaption price*

$$P(t, T_i, T_j)\, \mathbb{E}_{i,j}\left[\left(S(T_i, T_i, T_j) - \kappa\right)^+ \Big| \mathcal{F}_t\right]$$

can be approximated by

$$P(t, T_i, T_j) \mathrm{Bl}(\kappa, S(t, T_i, T_j), \tilde{\sigma}_{i,j}(t), 0, T_i - t), \qquad (11.13)$$

where

$$|\tilde{\sigma}_{i,j}(t)|^2 \qquad\qquad\qquad\qquad\qquad\qquad\qquad\qquad (11.14)$$

$$= \frac{1}{T_i - t} \sum_{l,k=i}^{j-1} \frac{\delta_k \delta_l v_{l+1}^{i,j}(t) v_{k+1}^{i,j}(t) L(t, T_l, T_{l+1}) L(t, T_k, T_{k+1})}{|S(t, T_i, T_j)|^2} \int_t^{T_i} \gamma_l(s)\gamma_k(s)\,ds.$$

Proof. We refer to Chapter 1 of [Schoenmakers (2005)] for a more rigorous treatment. Here we simply note that this approximation can be derived as follows:

$$dS(t, T_i, T_j) = d\left(\frac{1}{P(t, T_i, T_j)} \sum_{k=i}^{j-1} (T_{k+1} - T_k) P(t, T_{k+1}) L(t, T_k, T_{k+1})\right)$$

$$\simeq \frac{1}{P(t, T_i, T_j)} \sum_{k=i}^{j-1} (T_{k+1} - T_k) P(t, T_{k+1})\, dL(t, T_k, T_{k+1})$$

$$= \frac{1}{P(t, T_i, T_j)} \sum_{k=i}^{j-1} \delta_k P(t, T_{k+1}) L(t, T_k, T_{k+1}) \gamma_k(t)\, dB_t^{k+1}$$

$$= S(t, T_i, T_j) \sum_{k=i}^{j-1} \frac{\delta_k P(t, T_{k+1}) \gamma_k(t)}{S(t, T_i, T_j) P(t, T_i, T_j)} L(t, T_k, T_{k+1})\, dB_t^{k+1}$$

$$= S(t, T_i, T_j) \sum_{k=i}^{j-1} \delta_k v_{k+1}^{i,j}(t) \gamma_k(t) \frac{L(t, T_k, T_{k+1})}{S(t, T_i, T_j)}\, dB_t^{k+1},$$

hence

$$\sigma_{i,j}^2(t)dt \simeq \left| \frac{dS(t,T_i,T_j)}{S(t,T_i,T_j)} \right|^2$$

$$\simeq \frac{1}{S(t,T_i,T_j)^2} \sum_{k=i}^{j-1}\sum_{l=i}^{j-1} \delta_k \delta_l v_{k+1}^{i,j}(t) v_{l+1}^{i,j}(t) \gamma_k(t)\gamma_l(t) L(t,T_l,T_{l+1}) L(t,T_l,T_{l+1})dt,$$

which, in view of Relation (11.11), supports the claim that $w_k^{i,j}$ can be approximated by $\delta_k v_{k+1}^{i,j}$, see Chapter 1 of [Schoenmakers (2005)] for details.

The Black volatility

$$|\sigma_{i,j}(t)|^2 \qquad (11.15)$$

$$= \frac{1}{T_i - t} \sum_{l,k=i}^{j-1} \int_t^{T_i} \frac{\delta_l \delta_k v_{l+1}^{i,j}(s) v_{k+1}^{i,j}(s) L(s,T_l,T_{l+1}) L(s,T_k,T_{k+1})}{|S(t,T_i,T_j)|^2} \gamma_l(s)\gamma_k(s)ds,$$

$0 \le t \le T_i$, is approximated by

$$|\tilde{\sigma}_{i,j}(t)|^2 \qquad (11.16)$$

$$\simeq \frac{1}{T_i - t} \sum_{l,k=i}^{j-1} \frac{\delta_k \delta_l v_{l+1}^{i,j}(t) v_{k+1}^{i,j}(t) L(t,T_l,T_{l+1}) L(t,T_k,T_{k+1})}{|S(t,T_i,T_j)|^2} \int_t^{T_i} \gamma_l(s)\gamma_k(s)ds,$$

by "freezing" the random coefficients $v_{l+1}^{i,j}(t)$, $v_{k+1}^{i,j}(t)$, $L(t,T_l,T_{l+1})$, $L(t,T_k,T_{k+1})$ and $S(t,T_i,T_j)$ at time t. $\qquad\square$

This approximation amounts to saying that $S(t,T_i,T_j)$, $t \in [0,T_i]$, is an exponential martingale with volatiliy coefficient $\tilde{\sigma}_{i,j}(t)$ under $\mathbb{P}_{i,j}$. Note also that we have $\tilde{\sigma}_{i,i+1}(t) = \sigma_i(t)$, hence (11.13) is indeed an extension of (11.12).

11.4 Calibration of the BGM Model

Figure 11.2 shows an example of market data expressed in terms of swaption volatilities $\sigma_{i,j}^B(t)$ by inversion of the swaption approximation formula (11.13). Here, the time to maturity $T_i - t$ is in ordinate and the period $T_j - T_i$ is in abscissa. This type of data can be also expressed in the form of a graph where the index i refers to the time to maturity $T_i - t$ and the index j refers to the period $T_j - T_i$ as in Figure 11.3.

Vol Swaption At The Money

	1M	3M	6M	2M	YM	4M	5M	7M	OM	1DM	3YM
39	17,4	17	14,7	1Y,5	12,5	16,7	16	13,6	11,7	11,6	0,6
18	15,4	17	14,7	1Y,5	12,5	16,7	16	13,6	11,7	11,6	0,6
38	17,1	17,6Y	15,DY	1Y,0Y	12,0Y	12	16,3	13,YY	13	11,Y	0,2Y
68	17,4	17,5	15,6	14,3	1Y,3	12,3	16,2	13,7	13,3	11,5	0,4
48	17,5	17,1	14,7	1Y,5	12,7	16,0	16,6	13,5	13,3	11,7	0,5
08	17,6	15,Y	14,6	1Y,3Y	12,2	16,4	16,DY	13,YY	13,1	11,5Y	0,5
1M	15,0	14,0	1Y,7	12,7	12	16,6	13,7	13,2	13	11,5	0,5
3M	14,6	1Y,3	12,2	16,5	16,1	13,4	13,3	11,0	11,4	11,6	0,6
6M	1Y,3	12,3	16,2	13,7	13,2	13	11,5	11,Y	11,3	11	0,3
2M	12,2	16,3	13,4	13,1	11,5	11,Y	11,3	11	1D,7	1D,5	7,7
YM	16,2	13,2	11,0	11,Y	11,3	11	1D,7	1D,5	1D,Y	1D,2	7,4
4M	13,7Y	11,0Y	11,2Y	11	1D,5Y	1D,YY	1D,2	1D,3Y	1D,1	1D	7,6
5M	13,6	11,Y	11	1D,Y	1D,6	1D,1	1D	0,7	0,5	0,4	7
7M	11,05	11,16	1D,45	1D,3	1D	0,7	0,5	0,Y6	0,26	0,66	5,76
OM	11,46	1D,55	1D,66	0,0	0,5	0,Y	0,2	0,35	0,15	0,D5	5,45
1DM	11,6	1D,2	1D	0,4	0,2	0,3	0,1	0	7,0	7,7	5,Y
13M	1D,7	1D,D2	0,Y7	0,37	0,D3	7,03	7,54	7,44	7,Y4	7,24	5,67
1YM	1D,3	0,Y	0,1	7,7	7,4	7,Y	7,2	7,6	7,3	7,1	5,3
3DM	0,Y	7,7	7,Y	7,3	7	7	7	7	5,0	5,0	4,0
3YM	7,7	7,1	5,0	5,4	5,2	5,Y	5,4	5,5	5,4	5,5	4,4
6DM	7,1	5,2	5,6	5	4,7	5	5,3	5,2	5,6	5,Y	4,6

Fig. 11.2   Swaption volatilities.

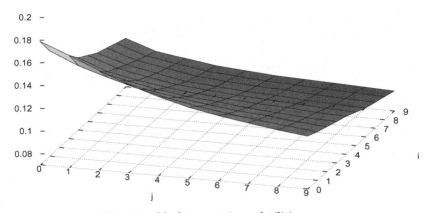

Fig. 11.3   Market swaption volatilities.

The goal of calibration is to estimate the volatility functions

$$\gamma_i(t) \in \mathbb{R}^d, \qquad 1 \le i \le n,$$

appearing in the BGM model (11.2) from the data of caps and swaptions prices observed on the market. This involves several computational and stability issues. Let

$$g_i(t) = |\gamma_i|(t), \qquad i = 1, \dots, n.$$

Using the [Rebonato (1996)] parameterization

$$g_i(t) = g_\infty + (1 + a(T_i - t) - g_\infty)e^{-b(T_i - t)},$$

$a, b, g_\infty > 0$, $i = 1, \ldots, n$, and equating

$$|\sigma_i^B(t)|^2 = \frac{1}{T_i - t} \int_t^{T_i} |\gamma_i|^2(s) ds$$

as in (11.8), one obtains from (11.15) an expression $\sigma_{i,j}(t, b, g_\infty)$ of $\sigma_{i,j}(t)$ as a function of b, g_∞, where a has been set equal to 0. Following [Schoenmakers (2002)] we minimize the mean square distance

$$\text{RMS}(b, g_\infty) := \sqrt{\frac{2}{(n-1)(n-2)} \sum_{i=1}^{k} \sum_{j=i+1}^{n} \left(\frac{\sigma_{i,j}^B(t) - \sigma_{i,j}(t)}{\sigma_{i,j}^B(t)} \right)^2},$$

where n is the number of tenor dates (in multiples of one year) and k is the maximum number of swaption maturities used in the calibration, with non-available data treated as zero in the sum. The data of discount factors and swap rates are interpolated with a fixed tenor $\delta = $ half year.

The volatilities computed in this way are given by the following graph, where the index i refers to $T_i - t$ and j refers to $T_j - T_i$:

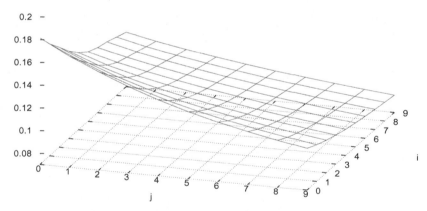

Fig. 11.4   Computed swaption volatilities.

The graph of Figure 11.5 allows us to compare the estimated and computed volatilities.

A sample of joint numerical estimation of the parameters (b, g_∞) is given in the table of Figure 11.6, where the maximum number k of swaption maturities used in each calibration is denoted by Nmat, see [Privault and Wei (2009)] for details. The total number of swaptions used is bounded by $nk - k(k+1)/2$.

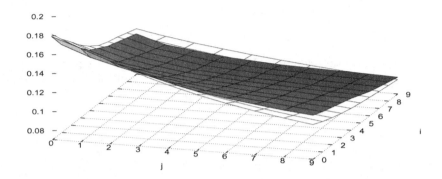

Fig. 11.5   Comparison graphs.

Nmat	#swaptions	b	g_∞	RMS
1	10	5.03	0.85	0.008
2	20	5.03	0.71	0.010
3	30	5.04	0.73	0.010
4	40	5.03	0.72	0.010
5	50	5.04	0.70	0.011
6	60	5.03	0.65	0.011
7	70	5.02	0.60	0.012
8	80	5.02	0.60	0.012
9	90	5.02	0.72	0.013
10	100	5.04	0.63	0.012
12	110	5.03	0.65	0.012
15	120	5.03	1.00	0.014

Fig. 11.6   Numerical results.

11.5   Exercises

Exercise 11.1. (Exercise 10.1 continued). Compute the price at time t:

$$\mathbb{E}\left[e^{-\int_t^{T_1} r_s ds} P(T_1, T_1, T_3)(S(T_1, T_1, T_3) - \kappa)^+ \Big| \mathcal{F}_t\right]$$
$$= P(t, T_1, T_3) \, \mathbb{E}_{1,3}\left[(S(T_1, T_1, T_3) - \kappa)^+ \mid \mathcal{F}_t\right],$$

of the swaption on $S(t, T_1, T_3)$ with strike κ, where $\mathbb{E}_{1,3}$ denotes the expectation under the forward swap measure $\mathbb{P}_{1,3}$ defined by

$$\frac{d\mathbb{P}_{1,3}}{d\mathbb{P}} = e^{-\int_0^{T_1} r_s ds} \frac{P(T_1, T_1, T_3)}{P(0, T_1, T_3)}.$$

You will need to use an approximation of $\sigma_{1,3}(s)$, for this it can be useful to "freeze" at time t all the random terms appearing in $\sigma_{1,3}(s)$, $s \geq t$.

Exercise 11.2. (Exercise 10.2 continued).

(1) Derive the stochastic differential equation satisfied by $P(t, T_1)$ and determine the process $\zeta_1(t)$ from the problem data.
(2) Show that $L(t, T_1, T_2)$ satisfies the stochastic differential equation

$$\frac{dL(t, T_1, T_2)}{L(t, T_1, T_2)} = \gamma dB_t - \gamma \zeta_2(t) dt. \qquad 0 \leq t \leq T_1, \qquad (11.17)$$

(3) Assume that $r_t = r > 0$ is a deterministic constant and that $\zeta_1(t) = 0$, $t \in \mathbb{R}_+$. Compute an approximation of the bond option price

$$P(t, T_1) \mathbb{E}_{\mathbb{P}} \left[(P(T_1, T_2) - K)^+ \big| \mathcal{F}_t \right]$$

as a function of $L(t, T_1, T_2)$. In order to derive an approximated price you may "freeze" the drift of $L(s, T_1, T_2)$ under $\mathbb{P}$, that is you may assume that (11.17) is written as

$$\frac{dL(s, T_1, T_2)}{L(s, T_1, T_2)} = \gamma dB_s - \gamma \zeta_2(t) ds, \qquad t \leq s \leq T_1.$$

The final result may be expressed as an integral over $\mathbb{R}$, whose explicit computation is not required.

Chapter 12

Appendix A: Mathematical Tools

This appendix surveys some basic results in probability and measure theory used in the lectures. It does not aim to completeness and the reader is referred to standard texts in probability such as [Jacod and Protter (2000)], [Protter (2005)] for more details.

In the sequel we work on a probability space $(\Omega, \mathcal{F}, \mathbb{P})$, and we denote by $\mathbb{E}$ the expectation under $\mathbb{P}$.

Measurability

Given a sequence $(Y_n)_{n \in \mathbb{N}}$ of random variables, a random variable F is said to be $\mathcal{F}_n$-measurable if it can be written as a function

$$F = f_n(Y_0, \ldots, Y_n)$$

of $Y_0, \ldots, Y_n$, where $f_n : \mathbb{R}^{n+1} \to \mathbb{R}$. This defines the natural filtration $(\mathcal{F}_n)_{n \geq -1}$ generated by $(Y_k)_{k \in \mathbb{N}}$, as

$$\mathcal{F}_n = \sigma(Y_0, \ldots, Y_n), \qquad n \geq 0,$$

and $\mathcal{F}_{-1} = \{\emptyset, \Omega\}$, where $\sigma(Y_0, \ldots, Y_n)$ is the smallest σ-algebra making $Y_0, \ldots, Y_n$ measurable.

A random variable X is said to be *integrable* if $\mathbb{E}[|X|] < \infty$.

Covariance and Correlation

The *covariance* of two random variables X and Y is defined as

$$\mathrm{Cov}(X, Y) = \mathbb{E}[(X - \mathbb{E}[X])(Y - \mathbb{E}[Y])],$$

with

$$\text{Cov}(X, X) = \mathbb{E}[(X - \mathbb{E}[X])^2]$$
$$= \mathbb{E}[X^2] - (\mathbb{E}[X])^2$$
$$= \text{Var}(X).$$

Moreover, for all $\alpha \in \mathbb{R}$ the variance satisfies the relation

$$\text{Var}(\alpha X) = \mathbb{E}[(\alpha X - \mathbb{E}[\alpha X])^2]$$
$$= \mathbb{E}[(\alpha X - \alpha \mathbb{E}[X])^2]$$
$$= \mathbb{E}[\alpha(X - \mathbb{E}[X])^2]$$
$$= \alpha^2 \mathbb{E}[(X - \mathbb{E}[X])^2]$$
$$= \alpha^2 \text{Var}(X). \tag{12.1}$$

The *correlation* of X and Y is the coefficient

$$c(X, Y) := \frac{\text{Cov}(X, Y)}{\sqrt{\text{Var}(X)}\sqrt{\text{Var}(Y)}}.$$

Clearly we have $c(X, Y) = 0$ when X is independent of Y, and $c(X, Y) = 1$ when $X = Y$. The conditional variance and covariance given a σ-algebra $\mathcal{G}$ can be defined similarly, with

$$\text{Cov}(X, Y|\mathcal{G}) = \mathbb{E}[(X - \mathbb{E}[X|\mathcal{G}])(Y - \mathbb{E}[Y|\mathcal{G}])|\mathcal{G}],$$

and

$$\text{Var}(X|\mathcal{G}) = \text{Cov}(X, X|\mathcal{G}) = \mathbb{E}[X^2|\mathcal{G}] - (\mathbb{E}[X|\mathcal{G}])^2.$$

Note that if Y is square-integrable and $\mathcal{G}$-measurable we have the relation

$$\text{Var}(X + Y|\mathcal{G}) = \mathbb{E}[(X + Y - \mathbb{E}[X + Y|\mathcal{G}])^2|\mathcal{G}]$$
$$= \mathbb{E}[(X - \mathbb{E}[X|\mathcal{G}])^2|\mathcal{G}]$$
$$= \text{Var}(X|\mathcal{G}).$$

An integrable random variable X is said to be *centered* if $\mathbb{E}[X] = 0$.

Gaussian Random Variables

A random variable X is Gaussian with mean μ and variance σ^2 if its characteristic function satisfies

$$\mathbb{E}[e^{i\alpha X}] = e^{i\alpha\mu - \alpha^2\sigma^2/2}, \qquad \alpha \in \mathbb{R},$$

i.e., in terms of Laplace transforms,

$$\mathbb{E}[e^{\alpha X}] = e^{\alpha\mu + \alpha^2\sigma^2/2}, \qquad \alpha \in \mathbb{R}. \tag{12.2}$$

From e.g. Corollary 16.1 of [Jacod and Protter (2000)] we have the following result.

Proposition 12.1. *Let $X_1, \ldots, X_n$ be an family of centered Gaussian variables which are assumed to be orthogonal to each other, i.e.*

$$\mathbb{E}[X_i X_j] = 0, \qquad 1 \le i \ne j \le n.$$

Then the random variables $X_1, \ldots, X_n$ are independent.

A couple (X, Y) of random variables is Gaussian with mean μ and covariance matrix Σ if its Laplace transform satisfies

$$\mathbb{E}[e^{i\langle X, u\rangle_{\mathbb{R}^2}}] = e^{i\langle \mu, u\rangle_{\mathbb{R}^2} - \frac{1}{2}\langle \Sigma u, u\rangle_{\mathbb{R}^2}}, \qquad u \in \mathbb{R}^2. \tag{12.3}$$

Finally, if $X_1, \ldots, X_n$ are independent Gaussian random variables with probability laws $\mathcal{N}(m_1, \sigma_1^2), \ldots, \mathcal{N}(m_n, \sigma_n^2)$ then then sum $X_1 + \cdots + X_n$ is a Gaussian random variable with probability law

$$\mathcal{N}(m_1 + \cdots + m_n, \sigma_1^2 + \cdots + \sigma_n^2).$$

Conditional Expectation

Consider $\mathcal{G}$ a sub σ-algebra of $\mathcal{F}$. The conditional expectation $\mathbb{E}[F \mid \mathcal{G}]$ of $F \in L^2(\Omega, \mathcal{F}, \mathbb{P})$ given $\mathcal{G}$ can be defined as the orthogonal projection of F on $L^2(\Omega, \mathcal{G}, \mathbb{P})$ for the inner product $\langle F, G \rangle := \mathbb{E}[FG]$. In other words, $\mathbb{E}[F \mid \mathcal{G}]$ is the only $\mathcal{G}$-measurable random variable that satisfies

$$\mathbb{E}[G(F - \mathbb{E}[F \mid \mathcal{G}])] = 0,$$

or

$$\mathbb{E}[GF] = \mathbb{E}[G\,\mathbb{E}[F \mid \mathcal{G}]], \tag{12.4}$$

for all G bounded and $\mathcal{G}$-measurable.

The conditional expectation has the following properties

a) $\mathbb{E}[F \mid \mathcal{G}] = \mathbb{E}[F]$ if F is independent of $\mathcal{G}$.

b) $\mathbb{E}[\mathbb{E}[F \mid \mathcal{F}] \mid \mathcal{G}] = \mathbb{E}[F \mid \mathcal{G}]$ if $\mathcal{G} \subset \mathcal{F}$.

c) $\mathbb{E}[GF \mid \mathcal{G}] = G\,\mathbb{E}[F \mid \mathcal{G}]$ if G is $\mathcal{G}$-measurable and sufficiently integrable.

d) We have

$$\mathbb{E}[f(X, Y) \mid \mathcal{F}] = \mathbb{E}[f(X, y)]_{y=Y} \tag{12.5}$$

if X, Y are independent and Y is $\mathcal{F}$-measurable.

Martingales in Discrete Time

Consider $(\mathcal{F}_n)_{n\in\mathbb{N}}$ an increasing family of sub σ-algebra of $\mathcal{F}$. A discrete time L^2-martingale with respect to $(\mathcal{F}_n)_{n\in\mathbb{N}}$ is a family $(M_n)_{n\in\mathbb{N}}$ of random variables such that

i) $M_n \in L^2(\Omega, \mathcal{F}_n, \mathbb{P})$, $n \in \mathbb{N}$,

ii) $\mathbb{E}[M_{n+1} \mid \mathcal{F}_n] = M_n$, $n \in \mathbb{N}$.

As an example, the process

$$(Y_0 + \cdots + Y_n)_{n\geq 0}$$

whose sequence $(Y_n)_{n\in\mathbb{N}}$ of increments satisfies

$$\mathbb{E}[Y_n \mid \mathcal{F}_{n-1}] = 0, \qquad n \in \mathbb{N}, \qquad\qquad (12.6)$$

is a martingale with respect to its own filtration defined as

$$\mathcal{F}_{-1} = \{\emptyset, \Omega\}$$

and

$$\mathcal{F}_n = \sigma(Y_0, \ldots, Y_n), \qquad n \geq 0.$$

In particular, Condition (12.6) is satisfied when the increments $(Y_n)_{n\in\mathbb{N}}$ are independent centered random variables.

Martingales in Continuous Time

Let $(\mathcal{F}_t)_{t\in\mathbb{R}_+}$ denote a continuous-time filtration, i.e. an increasing family of sub σ-algebras of $\mathcal{F}$. We assume that $(\mathcal{F}_t)_{t\in\mathbb{R}_+}$ is continuous on the right, i.e.

$$\mathcal{F}_t = \bigcap_{s>t} \mathcal{F}_s, \qquad t \in \mathbb{R}_+.$$

By a stochastic process we mean a family of random variables indexed by a time interval.

Definition 12.1. *A stochastic process $(M_t)_{t\in\mathbb{R}_+}$ such that $\mathbb{E}[|M_t|] < \infty$, $t \in \mathbb{R}_+$, is called an $\mathcal{F}_t$-martingale if*

$$\mathbb{E}[M_t|\mathcal{F}_s] = M_s, \qquad 0 \leq s \leq t.$$

A process $(X_t)_{t\in\mathbb{R}_+}$ is said to have independent increments if $X_t - X_s$ is independent of $\sigma(X_u \ : \ 0 \leq u \leq s), 0 \leq s < t$.

Proposition 12.2. *Every integrable process $(X_t)_{t\in\mathbb{R}_+}$ with centered independent increments is a martingale with respect to the filtration*

$$\mathcal{F}_t := \sigma(X_u \ : \ u \leq t), \quad t \in \mathbb{R}_+,$$

it generates, called the natural filtration.

Note that for all square-integrable random variable F the process $(\mathbb{E}[F|\mathcal{F}_t])_{t\in\mathbb{R}_+}$ is a martingale, due to the relation

$$\mathbb{E}[\mathbb{E}[F|\mathcal{F}_t]|\mathcal{F}_s] = \mathbb{E}[F|\mathcal{F}_s], \qquad 0 \leq s \leq t, \tag{12.7}$$

that follows from the so-called tower property (b) of the conditional expectation.

Markov Processes

Let $\mathcal{C}_0(\mathbb{R}^n)$ denote the class of continuous functions tending to 0 at infinity. Recall that f is said to tend to 0 at infinity if for all $\varepsilon > 0$ there exists a compact subset K of $\mathbb{R}^n$ such that $|f(x)| \leq \varepsilon$ for all $x \in \mathbb{R}^n \setminus K$.

Definition 12.2. *An $\mathbb{R}^n$-valued stochastic process, i.e. a family $(X_t)_{t\in\mathbb{R}_+}$ of $\mathbb{R}^n$-valued random variables on $(\Omega, \mathcal{F}, \mathbb{P})$, is a Markov process if for all $t \in \mathbb{R}_+$ the σ-fields*

$$\mathcal{F}_t^+ := \sigma(X_s \ : \ s \geq t)$$

and

$$\mathcal{F}_t := \sigma(X_s \ : \ 0 \leq s \leq t)$$

are conditionally independent given X_t.

This condition can be restated by saying that for all $A \in \mathcal{F}_t^+$ and $B \in \mathcal{F}_t$ we have

$$\mathbb{P}(A \cap B \mid X_t) = \mathbb{P}(A \mid X_t)\mathbb{P}(B \mid X_t),$$

cf. Chung [Chung (2002)]. This definition naturally entails that:

i) $(X_t)_{t\in\mathbb{R}_+}$ is adapted with respect to $(\mathcal{F}_t)_{t\in\mathbb{R}_+}$, i.e. X_t is $\mathcal{F}_t$-measurable, $t \in \mathbb{R}_+$, and

ii) X_u is conditionally independent of $\mathcal{F}_t$ given X_t, for all $u \geq t$, i.e.

$$\mathbb{E}[f(X_u) \mid \mathcal{F}_t] = \mathbb{E}[f(X_u) \mid X_t], \qquad 0 \leq t \leq u,$$

for any bounded measurable function f on $\mathbb{R}^n$.

In particular,

$$\mathbb{P}(X_u \in A \mid \mathcal{F}_t) = \mathbb{E}[\mathbf{1}_A(X_u) \mid \mathcal{F}_t] = \mathbb{E}[\mathbf{1}_A(X_u) \mid X_t] = \mathbb{P}(X_u \in A \mid X_t),$$

$A \in \mathcal{B}(\mathbb{R}^n)$. Processes with independent increments provide simple examples of Markov processes. Indeed, for all bounded measurable functions f, g we have

$$\begin{aligned}
&\mathbb{E}[f(X_{t_1}, \ldots, X_{t_n})g(X_{s_1}, \ldots, X_{s_n}) \mid X_t] \\
&= \mathbb{E}[f(X_{t_1} - X_t + x, \ldots, X_{t_n} - X_t + x) \\
&\quad g(X_{s_1} - X_t + x, \ldots, X_{s_n} - X_t + x)]_{x=X_t} \\
&= \mathbb{E}[f(X_{t_1} - X_t + x, \ldots, X_{t_n} - X_t + x)]_{x=X_t} \\
&\quad \mathbb{E}[g(X_{s_1} - X_t + x, \ldots, X_{s_n} - X_t + x)]_{x=X_t} \\
&= \mathbb{E}[f(X_{t_1}, \ldots, X_{t_n}) \mid X_t] \, \mathbb{E}[g(X_{s_1}, \ldots, X_{s_n}) \mid X_t],
\end{aligned}$$

$0 \leq s_1 < \cdots < s_n < t < t_1 < \cdots < t_n$.

A transition kernel is a mapping $\mathbb{P}(x, dy)$ such that

i) for every $x \in \mathbb{R}^n$, $A \longmapsto \mathbb{P}(x, A)$ is a probability measure, and

ii) for every $A \in \mathcal{B}(\mathbb{R}^n)$, the mapping $x \longmapsto \mathbb{P}(x, A)$ is a measurable function.

The transition kernel $\mu_{s,t}$ associated to $(X_t)_{t \in \mathbb{R}_+}$ is defined as

$$\mu_{s,t}(x, A) = \mathbb{P}(X_t \in A \mid X_s = x) \qquad 0 \leq s \leq t,$$

and we have

$$\mu_{s,t}(X_s, A) = \mathbb{P}(X_t \in A \mid X_s) = \mathbb{P}(X_t \in A \mid \mathcal{F}_s), \qquad 0 \leq s \leq t.$$

The transition operator $(P_{s,t})_{0 \leq s \leq t}$ associated to $(X_t)_{t \in \mathbb{R}_+}$ is defined as

$$P_{s,t}f(x) = \mathbb{E}[f(X_t) \mid X_s = x] = \int_{\mathbb{R}^n} f(y)\mu_{s,t}(x, dy), \qquad x \in \mathbb{R}^n.$$

Letting $p_{s,t}(x)$ denote the density of $X_t - X_s$ we have

$$\mu_{s,t}(x, A) = \int_A p_{s,t}(y - x)dy, \qquad A \in \mathcal{B}(\mathbb{R}^n),$$

and

$$P_{s,t}f(x) = \int_{\mathbb{R}^n} f(y)p_{s,t}(y-x)dy.$$

In the sequel we will assume that $(X_t)_{t\in\mathbb{R}_+}$ is time homogeneous, i.e. $\mu_{s,t}$ depends only on the difference $t-s$, and we will denote it by μ_{t-s}. In this case the family $(P_{0,t})_{t\in\mathbb{R}_+}$ is denoted by $(P_t)_{t\in\mathbb{R}_+}$ and defines a transition semigroup associated to $(X_t)_{t\in\mathbb{R}_+}$, with

$$P_t f(x) = \mathbb{E}[f(X_t) \mid X_0 = x] = \int_{\mathbb{R}^n} f(y)\mu_t(x, dy), \qquad x \in \mathbb{R}^n.$$

It satisfies the semigroup property

$$
\begin{aligned}
P_t P_s f(x) &= \mathbb{E}[P_s f(X_t) \mid X_0 = x] \\
&= \mathbb{E}[\mathbb{E}[f(X_{t+s}) \mid X_s] \mid X_0 = x]] \\
&= \mathbb{E}[\mathbb{E}[f(X_{t+s}) \mid \mathcal{F}_s] \mid X_0 = x]] \\
&= \mathbb{E}[f(X_{t+s}) \mid X_0 = x] \\
&= P_{t+s}f(x),
\end{aligned}
$$

which leads to the Chapman-Kolmogorov equation

$$\mu_{s+t}(x, A) = \mu_s * \mu_t(x, A) = \int_{\mathbb{R}^n} \mu_s(x, dy)\mu_t(y, A). \qquad (12.8)$$

By induction we obtain

$$\mathbb{P}_x((X_{t_1}, \ldots, X_{t_n}) \in B_1 \times \cdots \times B_n)$$
$$= \int_{B_1} \cdots \int_{B_n} \mu_{t_{n-1},t_n}(x_{n-1}, dx_n) \cdots \mu_{0,t_1}(x, dx_1),$$

for $0 < t_1 < \cdots < t_n$ and $B_1, \ldots, B_n$ Borel subsets of $\mathbb{R}^n$.

If $(X_t)_{t\in\mathbb{R}_+}$ is a homogeneous Markov processes with independent increments then the density $p_t(x)$ of X_t satisfies the convolution property

$$p_{s+t}(x) = \int_{\mathbb{R}^n} p_s(y-x)p_t(y)dy, \qquad x \in \mathbb{R}^n,$$

which is satisfied in particular by processes with stationary and independent increments such as Lévy processes. A typical example of a probability density satisfying such a convolution property is the Gaussian density, i.e.

$$p_t(x) = \frac{1}{(2\pi t)^{n/2}} \exp\left(-\frac{1}{2t}\|x\|_{\mathbb{R}^n}^2\right), \qquad x \in \mathbb{R}^n.$$

Chapter 13

Appendix B: Some Recent
Developments

In this appendix we list some recent issues investigated in the literature.

Infinite dimensional analysis

The modern mathematical modeling of long term interest rates relies largely
on functional analytic tools such as infinite dimensional Lie algebras and
manifolds, adding another level of technical difficulty in comparison with
standard equity models. Indeed, forward rates $F(t, T, S)$ can be reinterpreted as processes $t \longmapsto F(t, \cdot, \cdot)$ taking values in a function space of two
or more variables, thus their modeling makes a significant use of stochastic
processes taking values in (infinite-dimensional) function spaces. This approach has seen a considerable development in recent years, cf. e.g. [Björk
(2004)], [De Donno and Pratelli (2005)], [Ekeland and Taflin (2005)],
[Filipović and Teichmann (2004)], [Pratelli (2008)], [Da Prato (2004)].
For example, the HJM model considered in Chapter 6 can be written in
terms of stochastic evolution equations in Banach spaces, see [Carmona and
Tehranchi (2006)] and references therein for this approach. Related prob-
lems considered in the literature include the following items, see e.g. [Björk
(2004)] and references therein:

a) invariant manifolds: the determination of a function space containing
   the initial condition $r_0(\cdot)$, in which the forward rate process remains
   through its time evolution.

b) finite dimensional realizations: to find conditions for $f_t(\cdot)$ to remain in
   a finite dimensional manifold of interest rate curves.

c) consistency: to find conditions for $f_t(\cdot)$ to belong to one of the existing function spaces used for the modeling of forward interest rate curves, e.g. the Nelson-Siegel space.

d) hedging of maturity-related risks [Carmona and Tehranchi (2006)].

A negative answer has been given to point (c) above, see §3.5 of [Björk (2004)]. Actually from Relation (6.3) we already know that the Vasicek instantaneous forward rates $x \longmapsto f(t, t+x)$ live in a space which is different from both the Nelson-Siegel and the Svensson spaces.

Extended interest rate models

Both the Vasicek and CIR stochastic short rate models belong to the family of so-called affine models which has been extended in several directions. For example, quadratic models have been constructed using matrix-valued diffusion processes, cf. e.g. [Gourieroux and Sufana (2003)] and references therein. Other models preserving the positivity of interest rates have been proposed, cf. e.g. §16.4 of [James and Webber (2001)], using stochastic differential equations on Lie groups. In recent years the LIBOR market model has also been extended to include stochastic volatility, see [Andersen and Brotherton-Ratcliffe (2005)], [Piterbarg (2004)], and [Wu and Zhang (2006)]. LIBOR models with jumps have been introduced in [Glasserman and Kou (2003)] for marked point processes and in [Eberlein and Özkan (2005)] using Lévy processes.

Exotic and path-dependent options on interest rates

This type option is likely to become increasingly popular. Let us mention two examples.

Target Redemption Notes (TARN). The payment of such options is based on the time when the accumulated coupon received

$$\int_0^t (\kappa - F(s,T))^+ ds$$

reaches a given value, making it dependent on the whole path of forward spot rates $F(s,T)$. This type of option is already popular on Asian markets and is likely to gain more future development.

Range Accrual Notes (RAN). In this type of option, enhanced interest is accrued each day that the reference index stays within a predetermined range $[m, M]$, leading to a payoff of the form

$$\frac{1}{S-T} \int_T^S \mathbf{1}_{\{F(t,T)\in[m,M]\}} dt.$$

Other exotic options include snowball options and volatility bonds.

Sensitivity analysis and the Malliavin calculus

An important practical issue in mathematical finance is the fast numerical computation of price functions and of their derivatives (also called sensitivities) with respect to market model parameters. The computation of these sensitivities is useful for the hedging and risk management of financial claims, with speed of computation being a major concern. Fast numerical schemes for the computation of sensitivities of equity option prices to factors such as spot price, interest rate, or volatility parameters have been obtained via the stochastic calculus of variations (or Malliavin calculus) via formulas of the type

$$\frac{\partial}{\partial x} E[\phi(S_T) \mid S_0 = x] = E[\phi(S_T)\Lambda_T | S_0 = x], \tag{13.1}$$

where Λ_T is a stochastic weight computed using a stochastic gradient D in a direction u and its adjoint δ, cf. [Fournié et al. (1999)]. The stochastic calculus of variations is a flexible tool which has been developed not only on linear path spaces but also on Riemannian path spaces, including in the infinite dimensional case, cf. e.g. [Malliavin (1997)], [Malliavin and Thalmaier (2006)]. Applications of the stochastic calculus of variations in order to derive sensitivity formulas analogous to (13.1) on forward interest rate options have been developed in [Da Fonseca and Messaoud (2009)].

Longevity and mortality risk

Recently, stochastic models derived from interest rate modeling have been proposed for longevity and mortality risks, starting with [Milevsky and Promislow (2001)]. This approach has lead to the construction and pricing of longevity bonds and mortality derivatives, and the HJM framework has found new applications in this context.

Chapter 14

Solutions to the Exercises

Chapter 1

Exercise 1.1. We need to check the five properties of the Brownian motion:

(i) starts at 0 at time 0,

(ii) independence of increments,

(iii) almost sure continuity of trajectories,

(iv) stationarity of the increments,

(v) Gaussianity of the increments.

Checking conditions (i) to (iv) does not pose any particular problem since the time changes $t \longmapsto c+t$ and $t \longmapsto t/c^2$ are deterministic and continuous. Concerning (v), $B_{c+t} - B_c$ clearly has a centered Gaussian distribution with variance t, and the same property holds for cB_{t/c^2} since

$$\mathrm{Var}(cB_{t/c^2}) = c^2 \, \mathrm{Var}(B_{t/c^2}) = c^2 t/c^2 = t.$$

Exercise 1.2. We have $S_t = S_0 e^{\sigma B_t - \sigma^2 t/2 + \mu t}$.

Exercise 1.3. Looking for a solution of the form

$$X_t = a(t) \left(x_0 + \int_0^t b(s) dB_s \right)$$

175

yields

$$X_t = e^{-\alpha t} x_0 + \sigma \int_0^t e^{-\alpha(t-s)} dB_s, \qquad t \in \mathbb{R}_+.$$

Exercise 1.4. By the proposed identification we get $a'(t)/a(t) = t$ and $a(t)b(t) = e^{t^2/2}$, hence $a(t) = e^{t^2/2}$ and $b(t) = 1$, which yields $X_t = e^{t^2/2}(x_0 + B_t)$, $t \in \mathbb{R}_+$.

Exercise 1.5. Letting $X_t = \sqrt{Y_t}$ we have $dX_t = \mu X_t dt + \sigma dB_t$, hence

$$Y_t = \left(e^{\mu t} \sqrt{y_0} + \sigma \int_0^t e^{\mu(t-s)} dB_s \right)^2.$$

Exercise 1.6. Let $f \in L^2([0,T])$. We have

$$\mathbb{E}\left[e^{\int_0^T f(s)dB_s} \Big| \mathcal{F}_t \right] = \exp\left(\int_0^t f(s)dB_s + \frac{1}{2} \int_0^T |f(s)|^2 ds \right), \qquad 0 \le t \le T.$$

Exercise 1.7.
We have

$$\mathbb{E}\left[\exp\left(\beta \int_0^T B_t dB_t \right) \right] = \mathbb{E}\left[\exp\left(\beta (B_T^2 - T)/2 \right) \right]$$

$$= e^{-\beta T/2} \mathbb{E}\left[\exp\left(e^{\beta |B_T|^2/2} \right) \right]$$

$$= \frac{e^{-\beta T/2}}{\sqrt{2\pi T}} \int_{-\infty}^{\infty} e^{(\beta - \frac{1}{T}) \frac{x^2}{2}} dx$$

$$= \frac{e^{-\beta T/2}}{\sqrt{1 - \beta T}}$$

for all $\beta < 1/T$.

Exercise 1.8.

(1) We have

$$d(X_t^T/(T-t)) = \frac{dX_t^T}{T-t} + \frac{X_t^T}{|T-t|^2} dt = \sigma \frac{dB_t}{T-t},$$

hence by integration using the initial condition $X_0 = 0$ we have

$$\frac{X_t^T}{T-t} = \sigma \int_0^t \frac{1}{T-s} dB_s, \qquad 0 \le t \le T.$$

(2) We have

$$\mathbb{E}[X_t^T] = \sigma(T-t)\,\mathbb{E}\left[\int_0^t \frac{1}{T-s}dB_s\right] = 0.$$

(3) Using the Itô isometry we have

$$\mathrm{Var}[X_t^T] = \sigma^2|T-t|^2\,\mathrm{Var}\left[\int_0^t \frac{1}{T-s}dB_s\right]$$

$$= \sigma^2|T-t|^2 \int_0^t \frac{1}{|T-s|^2}ds$$

$$= \sigma^2|T-t|^2\left(\frac{1}{T-t}-\frac{1}{T}\right)$$

$$= \sigma^2(1-t/T).$$

(4) We have $\mathrm{Var}[X_T^T] = 0$ hence $X_T^T = \mathbb{E}[X_T^T] = 0$ by Question (2).

Chapter 2

Exercise 2.1.

(1) We have

$$X_t^\alpha = e^{\alpha t}X_0^\alpha + \int_0^t e^{\alpha(t-s)}dB_s,$$

and the Girsanov Theorem 2.1 yields

$$\frac{d\mathbb{Q}}{d\mathbb{P}} = \exp\left(-\alpha \int_0^T X_t^{(\alpha)}dB_t - \frac{\alpha^2}{2}\int_0^T |X_t^{(\alpha)}|^2dt\right).$$

(2) For all $\beta < 1/T$ we have

$$\mathbb{E}\left[\exp\left((\beta-\alpha)\int_0^T X_t^{(\alpha)}dX_t^{(\alpha)} + \frac{\alpha^2}{2}\int_0^T |X_t^{(\alpha)}|^2dt\right)\right]$$

$$= \mathbb{E}\left[\exp\left(\beta\int_0^T X_t^{(\alpha)}dX_t^{(\alpha)} - \alpha\int_0^T X_t^{(\alpha)}dB_t\right.\right.$$

$$\left.\left. -\alpha^2\int_0^T |X_t^{(\alpha)}|^2dt + \frac{\alpha^2}{2}\int_0^T |X_t^{(\alpha)}|^2dt\right)\right]$$

$$= \mathbb{E}\left[\exp\left(\beta\int_0^T X_t^{(\alpha)}dX_t^{(\alpha)} - \alpha\int_0^T X_t^{(\alpha)}dB_t - \frac{\alpha^2}{2}\int_0^T |X_t^{(\alpha)}|^2dt\right)\right]$$

$$= \mathbb{E}_{\mathbb{Q}}\left[\exp\left(\beta \int_0^T X_t^{(\alpha)} dX_t^{(\alpha)}\right)\right]$$

$$= \mathbb{E}\left[\exp\left(\beta \int_0^T B_t dB_t\right)\right]$$

$$= \frac{e^{-\beta T/2}}{\sqrt{1 - \beta T}},$$

using the fact that $\int_0^T B_t dB_t = (B_T^2 - T)/2$, where $B_T \simeq \mathcal{N}(0, T)$.

(3) Taking $\beta = \alpha$ yields

$$\mathbb{E}\left[\exp\left(\frac{\alpha^2}{2}\int_0^T |X_t^{(\alpha)}|^2 dt\right)\right] = \frac{e^{-\alpha T/2}}{\sqrt{1 - \alpha T}},$$

for all $\alpha < 1/T$.

Exercise 2.2.

(1) By the Girsanov Theorem 2.1, the probability $\mathbb{Q}$ is given by its density

$$\frac{d\mathbb{Q}}{d\mathbb{P}} = \exp\left(-\frac{\mu - r}{\sigma}B_T - \frac{|\mu - r|^2 T}{2\sigma^2}\right).$$

(2) For all $t \in [0, T]$ we have

$$C(t, S_t) = e^{-r(T-t)} S_t^2 \, \mathbb{E}_{\mathbb{Q}}\left[\frac{S_T^2}{S_t^2}\bigg| \mathcal{F}_t\right]$$

$$= e^{-r(T-t)} S_t^2 \, \mathbb{E}_{\mathbb{Q}}\left[e^{2\sigma(\tilde{B}_T - \tilde{B}_t) - \sigma^2(T-t) + 2r(T-t)}\bigg| \mathcal{F}_t\right]$$

$$= S_t^2 e^{(r+\sigma^2)(T-t)},$$

where $\tilde{B}_t = B_t + (\mu - r)t/\sigma$, $t \in [0, T]$, is a standard Brownian motion under $\mathbb{Q}$.

(3) For all $t \in [0, T]$ we have

$$\zeta_t = \frac{\partial C}{\partial x}(t, x)|_{x=S_t} = 2S_t e^{(r+\sigma^2)(T-t)},$$

and

$$\eta_t = \frac{C(t, S_t) - \zeta_t S_t}{A_t} = \frac{e^{-rt}}{A_0}(S_t^2 e^{(r+\sigma^2)(T-t)} - 2S_t^2 e^{(r+\sigma^2)(T-t)})$$

$$= -\frac{S_t^2}{A_0} e^{\sigma^2(T-t) + r(T-2t)}.$$

(4) For all $t \in [0, T_0]$ we have:

$$C(t) = e^{-r(T-t)} \, \mathbb{E}_{\mathbb{Q}} \left[\frac{S_T}{S_{T_0}} \Big| \mathcal{F}_t \right]$$

$$= e^{-r(T-t)} \, \mathbb{E}_{\mathbb{Q}} \left[\frac{S_T}{S_{T_0}} \right]$$

$$= e^{-r(T-t)} \, \mathbb{E}_{\mathbb{Q}} \left[e^{\sigma(\tilde{B}_T - \tilde{B}_{T_0}) - \sigma^2(T-T_0)/2 + r(T-T_0)} \right]$$

$$= e^{-r(T-t) + r(T-T_0)}$$

$$= e^{-r(T_0-t)},$$

and for $t \in]T_0, T]$:

$$C(t) = e^{-r(T-t)} \, \mathbb{E}_{\mathbb{Q}} \left[\frac{S_T}{S_{T_0}} \Big| \mathcal{F}_t \right]$$

$$= e^{-r(T-t)} \frac{S_t}{S_{T_0}} \, \mathbb{E}_{\mathbb{Q}} \left[\frac{S_T}{S_t} \Big| \mathcal{F}_t \right]$$

$$= e^{-r(T-t)} \frac{S_t}{S_{T_0}} \, \mathbb{E}_{\mathbb{Q}} \left[\frac{S_T}{S_t} \right]$$

$$= e^{-r(T-t)} \frac{S_t}{S_{T_0}} \, \mathbb{E}_{\mathbb{Q}} \left[e^{\sigma(\tilde{B}_T - \tilde{B}_t) - \sigma^2(T-t)/2 + r(T-t)} \right]$$

$$= \frac{S_t}{S_{T_0}}.$$

(5) For all $t \in [0, T_0]$ we have $\zeta_t = 0$ and $\eta_t = e^{-rT_0}/A_0$, and for $t \in]T_0, T]$ we have $\eta_t = 0$ and $\zeta_t = 1/S_{T_0}$. We have $d\zeta_t = d\eta_t = 0$ for $t \in [0, T_0]$ and $t \in]T_0, T]$, hence the portfolio is self-financing on these time intervals. On the other hand, at $t = T_0$ we also have $S_{T_0} d\zeta_{T_0} + A_{T_0} d\eta_{T_0} = S_{T_0} \times 1/S_{T_0} - A_{T_0} e^{-rT_0}/A_0 = 0$.

Exercise 2.3.

(1) We have

$$S_t = S_0 e^{\alpha t} + \sigma \int_0^t e^{\alpha(t-s)} dB_s.$$

(2) We have $\alpha_M = r$.

(3) After computing the conditional expectation

$$C(t, x) = e^{-r(T-t)} \exp\left(x e^{r(T-t)} + \frac{\sigma^2}{4r} (e^{2r(T-t)} - 1) \right).$$

(4) Here we need to note that the usual Black-Scholes argument applies and yields $\zeta_t = \partial C(t, S_t)/\partial x$, that is

$$\zeta_t = \exp\left(S_t e^{r(T-t)} + \frac{\sigma^2}{4r}(e^{2r(T-t)} - 1)\right).$$

Chapter 3

Exercise 3.1.

(1) We have $Y_t = e^{-at}y_0 + \frac{\theta}{a}(1 - e^{-at}) + \sigma \int_0^t e^{-a(t-s)}dB_s$.

(2) We have $dX_t = X_t\left(\theta + \frac{\sigma^2}{2} - a\log X_t\right)dt + \sigma X_t dB_t$.

(3) We have $r_t = \exp\left(e^{-at}\log r_0 + \frac{\theta}{a}(1 - e^{-at}) + \sigma \int_0^t e^{-a(t-s)}dB_s\right)$,
   with $\eta = \theta + \sigma^2/2$.

(4) We have
$$\mathbb{E}[r_t \mid \mathcal{F}_u]$$
$$= \exp\left(e^{-a(t-u)}\log r_u + \frac{\theta}{a}(1 - e^{-a(t-u)}) + \frac{\sigma^2}{4a}(1 - e^{-2a(t-u)})\right)$$

from (12.2), Property (b) of conditional expectations in Appendix A, and Proposition 1.3.

(5) We have
$$\mathbb{E}[r_t^2 \mid \mathcal{F}_u]$$
$$= \exp\left(2e^{-a(t-u)}\log r_u + 2\frac{\theta}{a}(1 - e^{-a(t-u)}) + \frac{\sigma^2}{a}(1 - e^{-2a(t-u)})\right)$$

and $\text{Var}[r_t|\mathcal{F}_u]$ can be computed using the above two expressions along with the identity $\text{Var}[r_t|\mathcal{F}_u] = \mathbb{E}[r_t^2|\mathcal{F}_u] - (\mathbb{E}[r_t|\mathcal{F}_u])^2$.

(6) We have $\lim_{t\to\infty} \mathbb{E}[r_t] = \exp\left(\frac{\theta}{a} + \frac{\sigma^2}{4a}\right)$ and

$$\lim_{t\to\infty} \text{Var}[r_t] = \exp\left(\frac{2\theta}{a} + \frac{\sigma^2}{a}\right) - \exp\left(\frac{2\theta}{a} + \frac{\sigma^2}{2a}\right).$$

Exercise 3.2.

(1) We have $r_t = r_0 + \int_0^t (\alpha - \beta r_s)ds + \sigma \int_0^t \sqrt{r_s}dB_s$.

(2) Using the fact that the expectation of the stochastic integral with respect to Brownian motion is zero, after taking expectations and then the derivative with respect to t on both sides of the above integral equation, we get the differential equation $u'(t) = \alpha - \beta u(t)$ with initial condition $u(s) = r_s$. This equation is solved as

$$u(t) = \frac{\alpha}{\beta}\left(1 - e^{-\beta(t-s)}\right) + r_s e^{-\beta(t-s)}, \qquad 0 \le s \le t.$$

(3) Apply Itô's formula to

$$r_t^2 = f\left(r_0 + \int_0^t (\alpha - \beta r_s)ds + \sigma \int_0^t \sqrt{r_s}dB_s\right),$$

with $f(x) = x^2$, to obtain $d(r_t)^2 = r_t(\sigma^2 + 2\alpha - 2\beta r_t)dt + 2r_t\sigma\sqrt{r_t}dB_t$.

(4) Taking again the conditional expectation and then derivatives on both sides of the above equation we get

$$v'(t) = (2\alpha + \sigma^2)u(t) - 2\beta v(t),$$

which implies

$$v'(t) = \frac{\alpha}{\beta}(2\alpha + \sigma^2) + (2\alpha + \sigma^2)\left(r_s - \frac{\alpha}{\beta}\right)e^{-\beta(t-s)} - 2\beta v(t).$$

The solution of this equation can written as the sum $v(t) = w(t)+p(t)$ of a solution $w(t) = ce^{-2\beta t}$ to the homogeneous equation $w'(t) = -2\beta w(t)$ and a particular solution $p(t)$ to the original equation. Searching for a particular solution of the form $p(t) = \zeta + \xi e^{-\beta(t-s)}$ yields

$$p(t) = \frac{\alpha(2\alpha + \sigma^2)}{2\beta^2} + \frac{2\alpha + \sigma^2}{\beta}\left(r_s - \frac{\alpha}{\beta}\right)e^{-\beta(t-s)},$$

hence

$$v(t) = ce^{-2\beta(t-s)} + \frac{\alpha(2\alpha + \sigma^2)}{2\beta^2} + \frac{2\alpha + \sigma^2}{\beta}\left(r_s - \frac{\alpha}{\beta}\right)e^{-\beta(t-s)},$$

and the initial condition $v(t) = r_s^2$ yields

$$v(t) = r_s^2 e^{-2\beta(t-s)} + \frac{\alpha(2\alpha + \sigma^2)}{2\beta^2}(1 - e^{-2\beta(t-s)})$$

$$+ \frac{2\alpha + \sigma^2}{\beta}\left(r_s - \frac{\alpha}{\beta}\right)(e^{-\beta(t-s)} - e^{-2\beta(t-s)}),$$

$0 \le s \le t$. In addition we can compute

$\mathrm{Var}[r_t|r_s] = \mathrm{Var}[r_t|\mathcal{F}_s] = \mathbb{E}[r_t^2|\mathcal{F}_s] - (\mathbb{E}[r_t|\mathcal{F}_s])^2$
$= v(t) - |u(t)|^2$

$$= r_s^2 e^{-2\beta(t-s)} + \frac{\alpha(2\alpha + \sigma^2)}{2\beta^2} (1 - e^{-2\beta(t-s)})$$

$$+ \frac{2\alpha + \sigma^2}{\beta} \left(r_s - \frac{\alpha}{\beta} \right) (e^{-\beta(t-s)} - e^{-2\beta(t-s)})$$

$$- \left(\frac{\alpha}{\beta} \left(1 - e^{-\beta(t-s)} \right) + r_s e^{-\beta(t-s)} \right)^2$$

$$= r_s^2 e^{-2\beta(t-s)} + \frac{\alpha(2\alpha + \sigma^2)}{2\beta^2} (1 - e^{-2\beta(t-s)})$$

$$+ \frac{2\alpha + \sigma^2}{\beta} \left(r_s - \frac{\alpha}{\beta} \right) (e^{-\beta(t-s)} - e^{-2\beta(t-s)})$$

$$- \left(\frac{\alpha}{\beta} \left(1 - e^{-\beta(t-s)} \right) \right)^2 - r_s^2 e^{-2\beta(t-s)} - 2\frac{\alpha}{\beta} \left(1 - e^{-\beta(t-s)} \right) r_s e^{-\beta(t-s)}$$

$$= \frac{\alpha(2\alpha + \sigma^2)}{2\beta^2} (1 - e^{-2\beta(t-s)})$$

$$+ \frac{2\alpha + \sigma^2}{\beta} \left(r_s - \frac{\alpha}{\beta} \right) (e^{-\beta(t-s)} - e^{-2\beta(t-s)})$$

$$- \left(\frac{\alpha}{\beta} \left(1 - e^{-\beta(t-s)} \right) \right)^2 - 2\frac{\alpha}{\beta} \left(1 - e^{-\beta(t-s)} \right) r_s e^{-\beta(t-s)}$$

$$= r_s \frac{\sigma^2}{\beta} \left(e^{-\beta(t-s)} - e^{-2\beta(t-s)} \right) + \frac{\alpha\sigma^2}{\beta^2} |1 - e^{-\beta(t-s)}|^2,$$

which satisfies the initial condition $\text{Var}[r_s | r_s] = 0$.

(5) The result follows by a direct application of the Itô formula.

(6) Again, from the Itô formula we have

$$dR_t = 2X_t dX_t + \frac{\sigma^2}{4} dt$$

$$= \left(\frac{\sigma^2}{4} - \beta X_t^2 \right) dt + \sigma X_t dB_t$$

$$= \left(\frac{\sigma^2}{4} - \beta R_t \right) dt + \sigma |X_t| dW_t$$

$$= \left(\frac{\sigma^2}{4} - \beta R_t \right) dt + \sigma \sqrt{R_t} dW_t, \qquad t > 0.$$

One could also show that $(W_t)_{t \in \mathbb{R}_+}$ is also a Brownian motion, thus providing an explicit solution to (3.4).

Chapter 4

Exercise 4.1.

(1) Consider the function $F(t, x)$ defined via

$$F(t, x) = \mathbb{E}\left[\exp\left(-\int_t^T r_s ds\right)\bigg| r_t = x\right], \qquad 0 \le t \le T.$$

We have

$$F(t, r_t) = F(t, r_0 + \theta t + \sigma W_t),$$

hence by standard arbitrage arguments the PDE satisfied by $F(t, x)$ is

$$-xF(t, x) + \frac{\partial F}{\partial t}(t, x) + \theta \frac{\partial F}{\partial x}(t, x) + \frac{1}{2}\sigma^2 \frac{\partial^2 F}{\partial x^2}(t, x) = 0,$$

with terminal condition $F(T, x) = 1$.

(2) We have

$$F(t, r_t) = \mathbb{E}\left[\exp\left(-\int_t^T (r_0 + \theta s + \sigma W_s)ds\right)\bigg|\mathcal{F}_t\right]$$

$$= \mathbb{E}\left[\exp\left(-\int_t^T (r_t + \theta(s - t) + \sigma(W_s - W_t))ds\right)\bigg|\mathcal{F}_t\right]$$

$$= \exp\left(-(T - t)r_t - \theta|T - t|^2/2\right)$$

$$\times \mathbb{E}\left[\exp\left(-\sigma\int_t^T (W_s - W_t)ds\right)\bigg|\mathcal{F}_t\right]$$

$$= \exp\left(-(T - t)r_t - \theta|T - t|^2/2\right)$$

$$\times \mathbb{E}\left[\exp\left(-\sigma\int_t^T (W_s - W_t)ds\right)\right]$$

$$= \exp\left(-(T - t)r_t - \theta|T - t|^2/2\right)$$

$$\times \mathbb{E}\left[\exp\left(-\sigma\int_0^{T-t} W_s ds\right)\right],$$

$0 \le t \le T$. Now,

$$\mathbb{E}\left[\exp\left(-\sigma\int_0^{T-t} W_s ds\right)\right] = \mathbb{E}\left[\exp\left(-\sigma\int_0^{T-t}\int_0^{T-t} 1_{\{u \le s\}} dW_u ds\right)\right]$$

$$= \mathbb{E}\left[\exp\left(-\sigma\int_0^{T-t}\int_0^{T-t} 1_{\{u \le s\}} ds dW_u\right)\right]$$

$$= \mathbb{E}\left[\exp\left(-\sigma\int_0^{T-t}\int_0^{T-t}1_{\{u\leq s\}}dsdW_u\right)\right]$$

$$= \mathbb{E}\left[\exp\left(-\sigma\int_0^{T-t}\int_u^{T-t}dsdW_u\right)\right]$$

$$= \mathbb{E}\left[\exp\left(-\sigma\int_0^{T-t}(T-t-u)dW_u\right)\right]$$

$$= \exp\left(\frac{\sigma^2}{2}\int_0^{T-t}|T-t-u|^2du\right)$$

$$= \exp\left(\frac{\sigma^2}{2}\int_0^{T-t}u^2du\right)$$

$$= \exp\left(\frac{\sigma^2}{6}(T-t)^3\right),$$

hence

$$F(t,r_t) = \exp\left(-r_t(T-t) - \theta|T-t|^2/2 + \sigma^2(T-t)^3/6\right),$$

and

$$F(t,x) = \exp\left(-x(T-t) - \theta|T-t|^2/2 + \sigma^2(T-t)^3/6\right).$$

(3) We have

$$\frac{\partial F}{\partial t}(t,x) = (x - \sigma^2|T-t|^2/2 + \theta(T-t))F(t,x),$$

$$\frac{\partial F}{\partial x}(t,x) = -(T-t)F(t,x),$$

and

$$\frac{\partial^2 F}{\partial x^2}(t,x) = |T-t|^2 F(t,x),$$

which shows by addition that $F(t,x)$ satisfy the PDE with the terminal condition $F(T,T) = 1$.

Exercise 4.2. From Proposition 4.1 and (1.9) the bond pricing PDE is given by

$$\begin{cases} \dfrac{\partial F}{\partial t}(t,x) = xF(t,x) - (\alpha - \beta x)\dfrac{\partial F}{\partial x}(t,x) - \dfrac{1}{2}\sigma^2 x^2 \dfrac{\partial^2 F}{\partial x^2}(t,x) \\[3mm] F(T,x) = 1. \end{cases}$$

When $\alpha = 0$ we search for a solution of the form

$$F(t, x) = e^{A(T-t)-xB(T-t)},$$

with $A(0) = B(0) = 0$, which implies

$$\begin{cases} A'(s) = 0 \\ \\ B'(s) + \beta B(s) + \frac{1}{2}\sigma^2 B^2(s) = 1, \end{cases}$$

hence in particular $A(s) = 0$, $s \in \mathbb{R}$, and $B(s)$ solves a Riccatti equation, whose solution is easily checked to be

$$B(s) = \frac{2(e^{\gamma s} - 1)}{2\gamma + (\beta + \gamma)(e^{\gamma s} - 1)},$$

with $\gamma = \sqrt{\beta^2 + 2\sigma^2}$.

Exercise 4.3.

(1) We have

$$P(t, T) = P(s, T) \exp\left(\int_s^t r_u \, du + \int_s^t \sigma_u^T \, dB_u - \frac{1}{2} \int_s^t |\sigma_u^T|^2 du \right),$$

$0 \le s \le t \le T$.

(2) We have

$$d\left(e^{-\int_0^t r_s \, ds} P(t, T) \right) = e^{-\int_0^t r_s \, ds} P(t, T) \sigma_t^T \, dB_t,$$

which gives a martingale after integration, from the properties of the Itô integral.

(3) By the martingale property of the previous question we have

$$\mathbb{E}\left[e^{-\int_0^T r_s \, ds} \Big| \mathcal{F}_t \right] = \mathbb{E}\left[P(T, T) e^{-\int_0^T r_s \, ds} \Big| \mathcal{F}_t \right]$$

$$= P(t, T) e^{-\int_0^t r_s \, ds}, \qquad 0 \le t \le T.$$

(4) By the previous question we have

$$P(t, T) = e^{\int_0^t r_s \, ds} \mathbb{E}\left[e^{-\int_0^T r_s \, ds} \Big| \mathcal{F}_t \right]$$

$$= \mathbb{E}\left[e^{\int_0^t r_s \, ds} e^{-\int_0^T r_s \, ds} \Big| \mathcal{F}_t \right]$$

$$= \mathbb{E}\left[e^{-\int_t^T r_s \, ds} \Big| \mathcal{F}_t \right], \qquad 0 \le t \le T,$$

since $e^{-\int_0^t r_s \, ds}$ is an $\mathcal{F}_t$-measurable random variable.

(5) We have

$$\frac{P(t,S)}{P(t,T)} = \frac{P(s,S)}{P(s,T)} \exp \left(\int_s^t (\sigma_u^S - \sigma_u^T) dB_u - \frac{1}{2} \int_s^t (|\sigma_u^S|^2 - |\sigma_u^T|^2) du \right)$$

$$= \frac{P(s,S)}{P(s,T)} \exp \left(\int_s^t (\sigma_u^S - \sigma_u^T) dB_u^T - \frac{1}{2} \int_s^t |\sigma_u^S - \sigma_u^T|^2 du \right),$$

$0 \leq t \leq T$, hence letting $s = t$ and $t = T$ in the above expression we have

$$P(T,S) = \frac{P(t,S)}{P(t,T)} \exp \left(\int_t^T (\sigma_s^S - \sigma_s^T) dB_s^T - \frac{1}{2} \int_t^T |\sigma_s^S - \sigma_s^T|^2 ds \right).$$

Exercise 4.4. We check that $P(T,T) = e^{X_T^T} = 1$.

Problem 4.5.

(1) We have

$$d \left(e^{-\int_0^t r_s ds} P(t,T) \right) = -e^{-\int_0^t r_s ds} P(t,T) r_t dt + e^{-\int_0^t r_s ds} dP(t,T)$$

$$= -e^{-\int_0^t r_s ds} P(t,T) r_t dt + e^{-\int_0^t r_s ds} dF(t,X_t)$$

$$= -e^{-\int_0^t r_s ds} P(t,T) r_t dt + e^{-\int_0^t r_s ds} \frac{\partial F}{\partial t}(t,X_t) dt$$

$$+ e^{-\int_0^t r_s ds} \frac{\partial F}{\partial x}(t,X_t) dX_t + \frac{1}{2} \sigma^2 e^{-\int_0^t r_s ds} \frac{\partial^2 F}{\partial x^2}(t,X_t) dt$$

$$= \sigma e^{-\int_0^t r_s ds} \frac{\partial F}{\partial x}(t,X_t) dB_t$$

$$e^{-\int_0^t r_s ds} \left(-r_t P(t,T) + \frac{\partial F}{\partial t}(t,X_t) - bX_t \frac{\partial F}{\partial x}(t,X_t) - \frac{\sigma^2}{2} \frac{\partial^2 F}{\partial x^2}(t,X_t) \right) dt.$$

Since

$$t \longmapsto e^{-\int_0^t r_s ds} P(t,T) = e^{-\int_0^t r_s ds} \mathbb{E}_{\mathbb{P}} \left[\exp \left(-\int_t^T r_s ds \right) \Big| \mathcal{F}_t \right]$$

$$= \mathbb{E}_{\mathbb{P}} \left[\exp \left(-\int_0^T r_s ds \right) \Big| \mathcal{F}_t \right],$$

is a martingale we get that

$$-r_t P(t,T) + \frac{\partial F}{\partial t}(t,X_t) - bX_t \frac{\partial F}{\partial x}(t,X_t) + \frac{1}{2} \sigma^2 \frac{\partial^2 F}{\partial x^2}(t,X_t) = 0,$$

and the PDE

$$-(r+x)F(t,x) + \frac{\partial F}{\partial t}(t,x) - bx \frac{\partial F}{\partial x}(t,x) + \frac{1}{2} \sigma^2 \frac{\partial^2 F}{\partial x^2}(t,x) = 0.$$

(2) We have

$$X_t = \sigma \int_0^t e^{-b(t-s)} dB_s, \qquad t \in \mathbb{R}_+.$$

(3) Integrating both sides of the stochastic differential equation defining $(X_t)_{t \in \mathbb{R}_+}$ we get

$$X_t = -b \int_0^t X_s ds + \sigma B_t,$$

hence

$$\int_0^t X_s ds = \frac{1}{b} (\sigma B_t - X_t)$$

$$= \frac{\sigma}{b} \left(B_t - \int_0^t e^{-b(t-s)} dB_s \right)$$

$$= \frac{\sigma}{b} \int_0^t (1 - e^{-b(t-s)}) dB_s.$$

(4) We have

$$\int_t^T X_s ds = \int_0^T X_s ds - \int_0^t X_s ds$$

$$= \frac{\sigma}{b} \int_0^T (1 - e^{-b(T-s)}) dB_s - \frac{\sigma}{b} \int_0^t (1 - e^{-b(t-s)}) dB_s$$

$$= -\frac{\sigma}{b} \left(\int_0^t (e^{-b(T-s)} - e^{-b(t-s)}) dB_s + \int_t^T (e^{-b(T-s)} - 1) dB_s \right).$$

(5) Applying Corollary 1.1, we have

$$\mathbb{E} \left[\int_t^T X_s ds \middle| \mathcal{F}_t \right]$$

$$= -\frac{\sigma}{b} \mathbb{E} \left[\int_0^t (e^{-b(T-s)} - e^{-b(t-s)}) dB_s + \int_t^T (e^{-b(T-s)} - 1) dB_s \middle| \mathcal{F}_t \right]$$

$$= -\frac{\sigma}{b} \mathbb{E} \left[\int_0^t (e^{-b(T-s)} - e^{-b(t-s)}) dB_s \middle| \mathcal{F}_t \right]$$

$$\quad - \frac{\sigma}{b} \mathbb{E} \left[\int_t^T (e^{-b(T-s)} - 1) dB_s \middle| \mathcal{F}_t \right]$$

$$= -\frac{\sigma}{b} \mathbb{E} \left[\int_0^t (e^{-b(T-s)} - e^{-b(t-s)}) dB_s \middle| \mathcal{F}_t \right]$$

$$= -\frac{\sigma}{b} \int_0^t (e^{-b(T-s)} - e^{-b(t-s)}) dB_s.$$

(6) We have

$$\mathbb{E}\left[\int_t^T X_s ds \middle| \mathcal{F}_t\right] = -\frac{\sigma}{b}\int_0^t (e^{-b(T-s)} - e^{-b(t-s)}) dB_s$$

$$= -\frac{\sigma}{b}(e^{-b(T-t)} - 1)\int_0^t e^{-b(t-s)} dB_s$$

$$= \frac{X_t}{b}(1 - e^{-b(T-t)}).$$

(7) From the properties of variance and conditional variance stated in Appendix A we have

$$\mathrm{Var}\left[\int_t^T X_s ds \middle| \mathcal{F}_t\right]$$

$$= \mathrm{Var}\left[-\frac{\sigma}{b}\left(\int_0^t (e^{-b(T-s)} - e^{-b(t-s)}) dB_s + \int_t^T (e^{-b(T-s)} - 1) dB_s\right) \middle| \mathcal{F}_t\right]$$

$$= \frac{\sigma^2}{b^2}\,\mathrm{Var}\left[\int_t^T (e^{-b(T-s)} - 1) dB_s \middle| \mathcal{F}_t\right]$$

$$= \frac{\sigma^2}{b^2}\,\mathrm{Var}\left[\int_t^T (e^{-b(T-s)} - 1) dB_s\right]$$

$$= \frac{\sigma^2}{b^2}\int_t^T |e^{-b(T-s)} - 1|^2 ds.$$

(8) Given $\mathcal{F}_t$, the random variable $\int_t^T X_s ds$ has a Gaussian distribution with conditional mean

$$\mathbb{E}\left[\int_t^T X_s ds \middle| \mathcal{F}_t\right] = \frac{X_t}{b}(1 - e^{-b(T-t)})$$

and conditional variance

$$\mathrm{Var}\left[\int_t^T X_s ds \middle| \mathcal{F}_t\right] = \frac{\sigma^2}{b^2}\int_t^T |e^{-b(T-s)} - 1|^2 ds.$$

(9) We have

$$P(t,T) = \mathbb{E}_{\mathbb{P}}\left[\exp\left(-\int_t^T r_s ds\right) \middle| \mathcal{F}_t\right]$$

$$= \exp\left(-r(T-t) - \mathbb{E}\left[\int_t^T X_s ds \middle| \mathcal{F}_t\right] + \frac{1}{2}\,\mathrm{Var}\left[\int_t^T X_s ds \middle| \mathcal{F}_t\right]\right)$$

$$= \exp\left(-r(T-t) - \frac{X_t}{b}(1 - e^{-b(T-t)}) + \frac{\sigma^2}{2b^2}\int_t^T |e^{-b(T-s)} - 1|^2 ds\right).$$

(10) We have

$$F(t,x) = \exp\left(-r(T-t) - \frac{x}{b}(1-e^{-b(T-t)}) + \frac{\sigma^2}{2b^2}\int_t^T |e^{-b(T-s)}-1|^2 ds\right),$$

hence

$$\frac{\partial F}{\partial t}(t,x) = \left(r + xe^{-b(T-t)} - \frac{\sigma^2}{2b^2}|e^{-b(T-t)}-1|^2\right)F(t,x),$$

$$\frac{\partial F}{\partial x}(t,x) = -\frac{1}{b}(1-e^{-b(T-t)})F(t,x),$$

and

$$\frac{\partial^2 F}{\partial x^2}(t,x) = \frac{1}{b^2}|1-e^{-b(T-t)}|^2 F(t,x),$$

which implies

$$-(r+x)F(t,x) + \frac{\partial F}{\partial t}(t,x) - bx\frac{\partial F}{\partial x}(t,x) + \frac{1}{2}\sigma^2\frac{\partial^2 F}{\partial x^2}(t,x) = 0.$$

Chapter 5

Exercise 5.1.

(1) We have

$$\log P(t,T) = -(T-t)r_0 - \theta(T^2-t^2)/2 + \sigma^2(T-t)^3/6 - (T-t)(r_t-\theta t-r_0)$$

and

$$\log P(t,S) = -(S-t)r_0 - \theta(S^2-t^2)/2 + \sigma^2(S-t)^3/6 - (S-t)(r_t-\theta t-r_0),$$

hence

$$\log P(t,T) - \log P(t,S)$$
$$= \theta(S^2-T^2)/2 + \sigma^2(T-t)^3/6 + (S-T)(r_t-\theta t) + \sigma^2(S-t)^3/6,$$

and

$$f(t,T,S) = \frac{1}{S-T}(\log P(t,T) - \log P(t,S))$$
$$= \frac{1}{S-T}(\theta(S^2-T^2)/2 + \sigma^2(T-t)^3/6$$
$$+ (S-T)(r_t-\theta t) + \sigma^2(S-t)^3/6).$$

(2) We have

$$
\begin{aligned}
f(t,T) &= -\frac{\partial}{\partial T}\log P(t,T) \\
&= r_0 + T\theta - \sigma^2|T-t|^2/2 + (r_t - \theta t - r_0) \\
&= (T-t)\theta - \sigma^2|T-t|^2/2 + r_t \\
&= f(0,T) + \sigma^2 t(T-t/2) + \sigma W_t. \tag{14.1}
\end{aligned}
$$

Exercise 5.2.

(1) We have

$$
\begin{aligned}
f(t,T,S) &= -\frac{\log P(t,S) - \log P(t,T)}{S-T} \\
&= -\frac{-r(S-t) - \frac{X_t}{b}(1 - e^{-b(S-t)}) + \frac{\sigma^2}{2b^2}\int_t^S |e^{-b(S-s)} - 1|^2 ds}{S-T} \\
&\quad + \frac{-r(T-t) - \frac{X_t}{b}(1 - e^{-b(T-t)}) + \frac{\sigma^2}{2b^2}\int_t^T |e^{-b(T-s)} - 1|^2 ds}{S-T} \\
&= r - \frac{X_t}{b}\frac{e^{-b(S-t)} - e^{-b(T-t)}}{S-T} \\
&\quad + \frac{\sigma^2}{2b^2}\int_t^T \frac{|e^{-b(T-s)} - 1|^2}{S-T}ds - \frac{\sigma^2}{2b^2}\int_t^S \frac{|e^{-b(S-s)} - 1|^2}{S-T}ds.
\end{aligned}
$$

(2) We have

$$
\begin{aligned}
f(t,T) &= \lim_{S \searrow T} f(t,T,S) \\
&= r + X_t e^{-b(T-t)} - \frac{\sigma^2}{b^2}\int_t^T e^{-b(T-s)}(e^{-b(T-s)} - 1)ds \\
&= r + X_t e^{-b(T-t)} - \frac{\sigma^2}{2b^2}|1 - e^{-b(T-t)}|^2. \tag{14.2}
\end{aligned}
$$

Exercise 5.3.

(1) We have

$$
\begin{aligned}
f(t,T,S) &= -\frac{1}{S-T}\left(X_t^S - X_t^T - \mu(S-T)\right) \\
&= \mu - \sigma\frac{1}{S-T}\left((S-t)\int_0^t \frac{1}{S-s}dB_s - (T-t)\int_0^t \frac{1}{T-s}dB_s\right) \\
&= \mu - \sigma\frac{1}{S-T}\int_0^t \left(\frac{S-t}{S-s} - \frac{T-t}{T-s}\right)dB_s
\end{aligned}
$$

$$= \mu - \sigma \frac{1}{S-T} \int_0^t \frac{(T-s)(S-t) - (T-t)(S-s)}{(S-s)(T-s)} dB_s$$

$$= \mu + \frac{\sigma}{S-T} \int_0^t \frac{(s-t)(S-T)}{(S-s)(T-s)} dB_s.$$

(2) We have

$$f(t,T) = \mu - \sigma \int_0^t \frac{t-s}{|T-s|^2} dB_s.$$

(3) We note that

$$\lim_{T \searrow t} f(t,T) = \mu - \sigma \int_0^t \frac{1}{t-s} dB_s$$

does not exist in $L^2(\Omega)$.

(4) By Itô's calculus we have

$$\frac{dP(t,T)}{P(t,T)} = \sigma dB_t + \frac{1}{2}\sigma^2 dt + \mu dt - \frac{X_t^T}{T-t} dt$$

$$= \sigma dB_t + \frac{1}{2}\sigma^2 dt - \frac{\log P(t,T)}{T-t} dt, \qquad 0 \le t \le T.$$

(5) Let

$$r_t^S = \mu + \frac{1}{2}\sigma^2 - \frac{X_t^S}{S-t}$$

$$= \mu + \frac{1}{2}\sigma^2 - \sigma \int_0^t \frac{1}{S-s} dB_s,$$

and apply the result of Exercise 4.3-(4).

Chapter 6

Exercise 6.1.

(1) By (14.1) we have

$$d_t f(t,T) = \sigma^2 (T-t) dt - \theta dt + dr_t = \sigma^2 (T-t) dt + \sigma dW_t.$$

(2) The HJM Condition (6.9) is satisfied since the drift of $d_t f(t,T)$ equals $\sigma \int_t^T \sigma ds = \sigma^2 (T-t)$.

Exercise 6.2.

(1) By (14.2) we have

$$d_t f(t,T) = e^{-b(T-t)} dX_t + bX_t e^{-b(T-t)} dt + \frac{\sigma^2}{b} e^{-b(T-t)}(1 - e^{-b(T-t)}) dt$$

$$= \frac{\sigma^2}{b} e^{-b(T-t)}(1 - e^{-b(T-t)}) dt + \sigma e^{-b(T-t)} dB_t.$$

(2) We have

$$\frac{\sigma^2}{b} e^{-b(T-t)}(1 - e^{-b(T-t)}) = \sigma e^{-b(T-t)} \int_t^T \sigma e^{-b(T-s)} ds,$$

which is the HJM absence of arbitrage Condition (6.9).

Chapter 7

Exercise 7.1.

(1) We have

$$dP(t,T) = dF(t,r_t) = r_t P(t,T) dt + \sigma \frac{\partial F}{\partial x}(t,r_t) dW_t,$$

where the remaining terms in factor of dt add up to zero from the martingale property of $t \longmapsto e^{-\int_0^t r_s ds} P(t,T)$. Hence

$$dP(t,T) = r_t P(t,T) dt - \sigma(T-t) F(t,x) dW_t,$$

and

$$\frac{dP(t,T)}{P(t,T)} = r_t dt - (T-t)\sigma dW_t.$$

(2) From Question 1 we have

$$d\left(\exp\left(-\int_0^t r_s ds\right) P(t,T)\right)$$

$$= -r_t \exp\left(-\int_0^t r_s ds\right) P(t,T) dt + \exp\left(-\int_0^t r_s ds\right) dP(t,T)$$

$$= \sigma \exp\left(-\int_0^t r_s ds\right) \frac{\partial F}{\partial x}(t,r_t) dW_t$$

$$= -\sigma(T-t) \exp\left(-\int_0^t r_s ds\right) F(t,r_t) dW_t.$$

(3) We have

$$\Psi(t) = \mathbb{E}_{\mathbb{P}}\left[\frac{d\tilde{\mathbb{P}}}{d\mathbb{P}}\bigg|\mathcal{F}_t\right] = \frac{P(t,T)}{P(0,T)}e^{-\int_0^t r_s ds}, \qquad 0 \le t \le T.$$

(4) We have

$$d\Psi(t) = \Psi(t)\zeta_t dW_t,$$

where

$$\zeta_t = -\sigma(T-t).$$

(5) We have

$$\Psi(t) = \exp\left(\int_0^t \zeta_s dW_s - \frac{1}{2}\int_0^t \zeta_s^2 ds\right),$$

hence

$$\mathbb{E}_{\mathbb{P}}\left[\frac{d\tilde{\mathbb{P}}}{d\mathbb{P}}\bigg|\mathcal{F}_T\right] = \Psi(T) = \exp\left(\int_0^T \zeta_s dW_s - \frac{1}{2}\int_0^T \zeta_s^2 ds\right).$$

(6) From the Girsanov Theorem 2.1,

$$d\hat{W}_t = \sigma(T-t)dt + dW_t$$

is a standard Brownian motion under $\tilde{\mathbb{P}}$, hence under $\tilde{\mathbb{P}}$ we have the dynamics

$$dr_r = \theta dt + \sigma dW_t = (\theta - \sigma^2(T-t))dt + \sigma d\hat{W}_t.$$

(7) We have

$$\mathbb{E}\left[e^{-\int_t^T r_s ds}(P(T,S)-K)^+\Big|\mathcal{F}_t\right]$$

$$= P(t,T)\tilde{\mathbb{E}}\left[(P(T,S)-K)^+\Big|\mathcal{F}_t\right]$$

$$= P(t,T)\tilde{\mathbb{E}}\left[(e^{-(S-T)r_0-\theta(S^2-T^2)/2+\sigma^2(S-T)^3/6-\sigma(S-T)W_T}-K)^+\Big|\mathcal{F}_t\right]$$

$$= P(t,T)$$

$$\times\tilde{\mathbb{E}}\left[\left(e^{-(S-T)r_0-\frac{\theta}{2}(S^2-T^2)+\frac{\sigma^2}{6}(S-T)^3-(S-T)\sigma\left(-\sigma\int_0^T(T-t)dt+\tilde{W}_T\right)}-K\right)^+\Big|\mathcal{F}_t\right]$$

$$= P(t,T)$$

$$\times\tilde{\mathbb{E}}\left[\left(e^{-(S-T)r_0-\frac{\theta}{2}(S^2-T^2)+\frac{\sigma^2}{6}(S-T)^3-(S-T)\sigma\left(-\sigma T^2/2+\tilde{W}_T\right)}-K\right)^+\Big|\mathcal{F}_t\right]$$

$$= P(t,T)\,\mathbb{E}[(e^{m+X}-K)^+|\mathcal{F}_t],$$

where

$$m = -(S-T)r_0 - \frac{\theta}{2}(S^2 - T^2) + \frac{\sigma^2}{6}(S-T)^3$$
$$-(S-T)\sigma\left(-\sigma T^2/2 + \tilde{W}_t\right)$$
$$= -(S-T)r_0 - \frac{\theta}{2}(S^2 - T^2) + \frac{\sigma^2}{6}(S-T)^3$$
$$+(S-T)\sigma\left(\sigma T^2/2 + \sigma t^2/2 - \sigma tT - W_t\right),$$

and X is centered Gaussian with conditional variance

$$v^2 = |S-T|^2\sigma^2(T-t)$$

Hence

$$\mathbb{E}\left[e^{-\int_t^T r_s ds}(P(T,S) - K)^+ \Big| \mathcal{F}_t\right]$$
$$= P(t,T)e^{m+\frac{v^2}{2}}\Phi(v + (m - \log K)/v) - KP(t,T)\Phi((m - \log K)/v),$$

where

$$\Phi(z) = \int_{-\infty}^z e^{-y^2/2}\frac{dy}{\sqrt{2\pi}}, \qquad z \in \mathbb{R}.$$

Now,

$$t \longmapsto \frac{P(t,S)}{P(t,T)}$$

is a martingale under $\tilde{\mathbb{P}}$ (cf. Proposition 7.2), hence

$$\mathbb{E}_{\tilde{\mathbb{P}}}\left[P(T,S)\Big|\mathcal{F}_t\right] = \mathbb{E}_{\tilde{\mathbb{P}}}\left[\frac{P(T,S)}{P(T,T)}\Big|\mathcal{F}_t\right]$$
$$= \mathbb{E}_{\tilde{\mathbb{P}}}\left[\frac{P(T,S)}{P(T,T)}\Big|\mathcal{F}_t\right]$$
$$= \frac{P(t,S)}{P(t,T)},$$

and

$$\frac{P(t,S)}{P(t,T)} = \mathbb{E}_{\tilde{\mathbb{P}}}\left[P(T,S)\Big|\mathcal{F}_t\right] = \mathbb{E}[e^{m+X}|\mathcal{F}_t],$$

where X is a centered Gaussian random variable with variance v^2, thus

$$m + \frac{1}{2}v^2 = \log\frac{P(t,S)}{P(t,T)}. \qquad (14.3)$$

As a consequence,

$$\mathbb{E}\left[e^{-\int_t^T r_s ds}(P(T,S) - K)^+ \Big| \mathcal{F}_t\right]$$

$$= P(t,S)\Phi\left(\frac{v}{2} + \frac{1}{v}\log\frac{P(t,S)}{KP(t,T)}\right) - KP(t,T)\Phi\left(-\frac{v}{2} + \frac{1}{v}\log\frac{P(t,S)}{KP(t,T)}\right).$$

Note that Relation (14.3) can be obtained through "brute force calculation" instead of using a martingale argument: we have

$$\log P(t,T) - \log P(t,S)$$

$$= -(T - t)r_0 - \theta(T^2 - t^2)/2 + \sigma^2(T - t)^3/6 - \sigma(T - t)W_t$$

$$\quad - \left(-(S - t)r_0 - \theta(S^2 - t^2)/2 + \sigma^2(S - t)^3/6 - \sigma(S - t)W_t\right)$$

$$= -(T - S)r_0 - \theta(T^2 - S^2)/2 - \sigma(T - S)W_t$$

$$\quad + \sigma^2((T - t)^3 - (S - t)^3)/6,$$

hence

$$m = \log P(t,S) - \log P(t,T) + \sigma^2((T - t)^3 - (S - t)^3)/6 + \frac{\sigma^2}{6}(S - T)^3$$

$$\quad + (S - T)\sigma^2\left(T^2/2 + t^2/2\right) - \sigma^2 tT(S - T)$$

$$= \log P(t,S) - \log P(t,T) + \sigma^2(T^3 - 3tT^2 + 3Tt^2 - S^3 + 3tS^2 - 3St^2)/6$$

$$\quad + \frac{\sigma^2}{6}(S - T)^3 + (S - T)\sigma^2\left(T^2/2 + t^2/2\right) - \sigma^2 tT(S - T)$$

$$= \log P(t,S) - \log P(t,T) + \sigma^2(-3tT^2 + 3Tt^2 + 3tS^2 - 3St^2)/6$$

$$\quad + \frac{\sigma^2}{6}(-3TS^2 + 3ST^2) + (S - T)\sigma^2\left(T^2/2 + t^2/2\right) - \sigma^2 tT(S - T),$$

and

$$m + \frac{v^2}{2} = \log P(t,S) - \log P(t,T) + \sigma^2(-3tT^2 + 3Tt^2 + 3tS^2 - 3St^2)/6$$

$$\quad + \frac{\sigma^2}{6}(-3TS^2 + 3ST^2) + (S - T)\sigma^2\left(T^2/2 + t^2/2\right)$$

$$\quad + \frac{1}{2}\sigma^2(S^2 - 2ST + T^2)(T - t) - \sigma^2 tT(S - T)$$

$$= \log P(t,S) - \log P(t,T).$$

Exercise 7.2.

(1) From the result of Exercise 4.5-(9) we have

$$\mathbb{E}_{\mathbb{P}}\left[\frac{d\tilde{\mathbb{P}}}{d\mathbb{P}}\Big| \mathcal{F}_t\right]$$

$$= e^{-\int_0^t r_s ds}\frac{P(t,T)}{P(0,T)}$$

$$= \exp\left(-\int_0^t r_s ds + rt - \frac{\sigma^2}{2b^2}\int_0^t |e^{-b(T-s)} - 1|^2 ds - \frac{X_t}{b}(1 - e^{-b(T-t)})\right)$$

$$= \exp\left(-\int_0^t X_s ds - \frac{\sigma^2}{2b^2}\int_0^t |e^{-b(T-s)} - 1|^2 ds - \frac{X_t}{b}(1 - e^{-b(T-t)})\right)$$

$$= \exp\left(\frac{1}{b}(X_t - \sigma B_t) - \frac{\sigma^2}{2b^2}\int_0^t |e^{-b(T-s)} - 1|^2 ds - \frac{X_t}{b}(1 - e^{-b(T-t)})\right)$$

$$= \exp\left(-\frac{\sigma}{b}B_t - \frac{\sigma^2}{2b^2}\int_0^t |e^{-b(T-s)} - 1|^2 ds + \frac{X_t}{b}e^{-b(T-t)}\right)$$

$$= \exp\left(-\frac{\sigma}{b}B_t - \frac{\sigma^2}{2b^2}\int_0^t |e^{-b(T-s)} - 1|^2 ds + \frac{\sigma}{b}\int_0^t e^{-b(T-s)} dB_s\right)$$

$$= \exp\left(-\frac{\sigma}{b}\int_0^t (1 - e^{-b(T-s)}) dB_s - \frac{\sigma^2}{2b^2}\int_0^t |e^{-b(T-s)} - 1|^2 ds\right),$$

and in particular for $t = T$ we get

$$\frac{d\tilde{\mathbb{P}}}{d\mathbb{P}} = \exp\left(-\frac{\sigma}{b}\int_0^T (1 - e^{-b(T-s)}) dB_s - \frac{\sigma^2}{2b^2}\int_0^T |e^{-b(T-s)} - 1|^2 ds\right).$$

(2) From the Girsanov Theorem 2.1,

$$\hat{B}_t := B_t + \frac{\sigma}{b}\int_0^t (1 - e^{-b(T-s)}) ds, \qquad 0 \le t \le T,$$

is a standard Brownian motion under the forward measure $\tilde{\mathbb{P}}$ and we have

$$dr_t = dX_t = -bX_t dt + \sigma dB_t = -bX_t dt - \frac{\sigma^2}{b}(1 - e^{-b(T-t)}) dt + \sigma d\hat{B}_t.$$

(3) When $b = 0$ we have

$$P(t,T) = \exp\left(-r(T - t) - (T - t)X_t + \frac{\sigma^2}{2}\int_t^T |T - s|^2 ds\right)$$

and

$$X_T = -\sigma^2 \int_0^T (T - s) ds + \sigma \hat{B}_T = -\frac{\sigma^2}{2}T^2 + \sigma \hat{B}_T,$$

hence

$$\mathbb{E}_{\mathbb{P}}\left[e^{-\int_0^T r_s ds}(P(T,S) - K)^+\right] = P(0,T)\,\mathbb{E}_{\tilde{\mathbb{P}}}\left[(P(T,S) - K)^+\right]$$

$$= P(0,T)\,\mathbb{E}_{\tilde{\mathbb{P}}}\left[\left(e^{-r(S-T) - (S-T)X_T + \frac{\sigma^2}{2}\int_T^S |S-s|^2 ds} - K\right)^+\right]$$

$$= P(0,T) \, \mathbb{E}_{\tilde{\mathbb{P}}} \left[\left(e^{-r(S-T)-(S-T)\left(-\frac{\sigma^2}{2}T^2+\sigma \hat{B}_T\right)+\frac{\sigma^2}{2}\int_T^S |S-s|^2 ds} - K \right)^+ \right]$$

$$= P(0,T) \, \mathbb{E}_{\tilde{\mathbb{P}}} \left[\left(e^{-r(S-T)+\frac{\sigma^2}{2}(S-T)T^2+\frac{\sigma^2}{6}(S-T)^3-(S-T)\sigma \hat{B}_T} - K \right)^+ \right].$$

From the relations

$$\mathbb{E}[(e^{m+X}-K)^+] = e^{m+\frac{v^2}{2}} \Phi(v+(m-\log K)/v) - K\Phi((m-\log K)/v),$$

where $m = -r(S-T) + \frac{\sigma^2}{2}(S-T)T^2 + \frac{\sigma^2}{6}(S-T)^3$, X is a centered Gaussian random variable with variance $v^2 = \sigma^2 T|S-T|^2$, and

$$\Phi(z) = \int_{-\infty}^{z} e^{-y^2/2} \frac{dy}{\sqrt{2\pi}}, \qquad z \in \mathbb{R},$$

and

$$-v^2/2 + \log(P(0,S)/P(0,T)) = -\frac{1}{2}\sigma^2 T|S-T|^2 - rS + \frac{\sigma^2}{6}S^3$$
$$- \left(-rT + \frac{\sigma^2}{6}T^3 \right)$$
$$= m, \qquad\qquad (14.4)$$

we finally obtain

$$\mathbb{E}_{\mathbb{P}} \left[e^{-\int_0^T r_s \, ds} (P(T,S) - K)^+ \right]$$

$$= P(0,S)\Phi \left(\frac{1}{v} \log \frac{P(0,S)}{KP(0,T)} + \frac{v}{2} \right) - KP(0,T)\Phi \left(\frac{1}{v} \log \frac{P(0,S)}{KP(0,T)} - \frac{v}{2} \right).$$

On the other hand, Relation (14.4) can be independently recovered from the fact that

$$t \longmapsto \frac{P(t,S)}{P(t,T)}$$

is a martingale under $\tilde{\mathbb{P}}$ (cf. Chapter 7). Hence

$$\mathbb{E}_{\tilde{\mathbb{P}}} \left[P(T,S) \Big| \mathcal{F}_t \right] = \mathbb{E}_{\tilde{\mathbb{P}}} \left[\frac{P(T,S)}{P(T,T)} \Big| \mathcal{F}_t \right] = \frac{P(t,S)}{P(t,T)},$$

and

$$\frac{P(0,S)}{P(0,T)} = \mathbb{E}_{\tilde{\mathbb{P}}} \left[P(T,S) \Big| \mathcal{F}_0 \right] = \mathbb{E}[(e^{m+X}-K)^+],$$

where X is a centered Gaussian random variable with variance v^2, hence

$$m + \frac{1}{2}v^2 = \log \frac{P(0,S)}{P(0,T)}.$$

Exercise 7.3. Recall that the forward measure is defined by

$$\frac{d\mathbb{P}_{i|\mathcal{F}_t}}{d\mathbb{P}_{|\mathcal{F}_t}} = \frac{e^{-\int_t^{T_i} r_s ds}}{P(t, T_i)}, \qquad 0 \le t \le T_i, \quad i = 1, 2,$$

with

$$\mathbb{E}\left[\frac{d\mathbb{P}_i}{d\mathbb{P}}\bigg|\mathcal{F}_t\right] = \frac{P(t, T_i)}{P(0, T_i)} e^{-\int_0^t r_s ds}, \qquad 0 \le t \le T_i, \quad i = 1, 2.$$

The forward swap measure is defined by

$$\frac{d\mathbb{P}_{1,2|\mathcal{F}_t}}{d\mathbb{P}_{|\mathcal{F}_t}} = \frac{P(T_1, T_2)}{P(t, T_2)} e^{-\int_t^{T_1} r_s ds}, \qquad 0 \le t \le T_1,$$

with

$$\mathbb{E}\left[\frac{d\mathbb{P}_{1,2}}{d\mathbb{P}}\bigg|\mathcal{F}_t\right] = \frac{P(t, T_2)}{P(0, T_2)} e^{-\int_0^t r_s ds}, \qquad 0 \le t \le T_1.$$

In particular,

$$\mathbb{E}\left[\frac{d\mathbb{P}_{1,2}}{d\mathbb{P}}\bigg|\mathcal{F}_{T_1}\right] = \mathbb{E}\left[\frac{d\mathbb{P}_2}{d\mathbb{P}}\bigg|\mathcal{F}_{T_1}\right] = \frac{P(T_1, T_2)}{P(0, T_2)} e^{-\int_0^{T_1} r_s ds},$$

which means that

$$\mathbb{E}\left[e^{-\int_0^{T_2} r_s ds} F\right] = P(0, T_2) \mathbb{E}\left[\frac{P(T_1, T_2)}{P(0, T_2)} e^{-\int_0^{T_1} r_s ds} F\right]$$

$$= P(0, T_2) \mathbb{E}_2[F]$$

$$= P(0, T_2) \mathbb{E}_{1,2}[F],$$

for all F integrable and $\mathcal{F}_{T_1}$-measurable. Moreover,

$$dB_t^i := dB_t - \zeta_t^i dt$$

is a standard Brownian motion under $\mathbb{P}_i$, $i = 1, 2$. In addition, $(B_t^2)_{t \in [0, T_1]}$ is also a standard Brownian motion until time T_1 under $\mathbb{P}_{1,2}$. We have

$$\mathbb{E}_{1,2}\left[\frac{d\mathbb{P}_k}{d\mathbb{P}_{1,2}}\bigg|\mathcal{F}_t\right] = \frac{P(0, T_2)}{P(0, T_k)} \frac{P(t, T_k)}{P(t, T_2)} \qquad 0 \le t \le T_1, \quad k = 1, 2,$$

and the process

$$t \longmapsto \frac{P(t, T_1)}{P(t, T_2)}, \qquad 0 \le t \le T_1,$$

is an $\mathcal{F}_t$-martingale under $\mathbb{P}_2$ and under $\mathbb{P}_{1,2}$, while

$$t \longmapsto \frac{P(t, T_2)}{P(t, T_1)}, \qquad 0 \le t \le T_1,$$

is an $\mathcal{F}_t$-martingale under $\mathbb{P}_1$.

(1) We have

$$\frac{dP(t, T_i)}{P(t, T_i)} = r_t dt + \zeta_t^i dB_t, \qquad i = 1, 2,$$

and

$$P(T, T_i) = P(t, T_i) \exp\left(\int_t^T r_s ds + \int_t^T \zeta_s^i dB_s - \frac{1}{2} \int_t^T |\zeta_s^i|^2 ds\right),$$

$0 \le t \le T \le T_i$, $i = 1, 2$, hence

$$\log P(T, T_i) = \log P(t, T_i) + \int_t^T r_s ds + \int_t^T \zeta_s^i dB_s - \frac{1}{2} \int_t^T |\zeta_s^i|^2 ds,$$

$0 \le t \le T \le T_i$, $i = 1, 2$, and

$$d \log P(t, T_i) = r_t dt + \zeta_t^i dB_t - \frac{1}{2} |\zeta_t^i|^2 dt, \qquad i = 1, 2.$$

In the Vasicek model

$$dr_t = -b r_t dt + \sigma dB_t,$$

where $(B_t)_{t \in \mathbb{R}_+}$ is a standard Brownian motion under $\mathbb{P}$, we have

$$\zeta_t^i = -\frac{\sigma}{b}(1 - e^{-b(T_i - t)}), \qquad 0 \le t \le T_i, \quad i = 1, 2.$$

Let

$$dB_t^i = dB_t - \zeta_t^i dt,$$

which is a standard Brownian motion under $\mathbb{P}_i$, $i = 1, 2$.

We also have

$$\frac{P(T, T_1)}{P(T, T_2)} = \frac{P(t, T_1)}{P(t, T_2)} \exp\left(\int_t^T (\zeta_s^1 - \zeta_s^2) dB_s - \frac{1}{2} \int_t^T (|\zeta_s^1|^2 - |\zeta_s^2|^2) ds\right)$$

$$= \frac{P(t, T_1)}{P(t, T_2)} \exp\left(\int_t^T (\zeta_s^1 - \zeta_s^2) dB_s^{(2)} - \frac{1}{2} \int_t^T |\zeta_s^1 - \zeta_s^2|^2 ds\right),$$

which is an $\mathcal{F}_t$-martingale under $\mathbb{P}_2$ and under $\mathbb{P}_{1,2}$, and

$$\frac{P(T, T_2)}{P(T, T_1)} = \frac{P(t, T_2)}{P(t, T_1)} \exp\left(-\int_t^T (\zeta_s^1 - \zeta_s^2) dB_s^{(1)} - \frac{1}{2} \int_t^T |\zeta_s^1 - \zeta_s^2|^2 ds\right),$$

which is an $\mathcal{F}_t$-martingale under $\mathbb{P}_1$.

(2) We have

$$f(t, T_1, T_2) = -\frac{1}{T_2 - T_1}(\log P(t, T_2) - \log P(t, T_1))$$

and in the Vasicek case,

$$f(t, T_1, T_2) = -\frac{\sigma^2}{2b}$$
$$-\frac{1}{T_2 - T_1}\left(\left(\frac{r_t}{b} + \frac{\sigma^2}{b^3}\right)\left(e^{-b(T_2-t)} - e^{-b(T_1-t)}\right)\right.$$
$$\left.-\frac{\sigma^2}{4b^3}\left(e^{-2b(T_2-t)} - e^{-2b(T_1-t)}\right)\right).$$

(3) We have

$$df(t, T_1, T_2) = -\frac{1}{T_2 - T_1}d\log\left(P(t, T_2)/P(t, T_1)\right)$$
$$= -\frac{1}{T_2 - T_1}\left((\zeta_t^2 - \zeta_t^1)dB_t - \frac{1}{2}(|\zeta_t^2|^2 - |\zeta_t^1|^2)dt\right)$$
$$= -\frac{1}{T_2 - T_1}\left((\zeta_t^2 - \zeta_t^1)(dB_t^{(2)} + \zeta_t^2 dt) - \frac{1}{2}(|\zeta_t^2|^2 - |\zeta_t^1|^2)dt\right)$$
$$= -\frac{1}{T_2 - T_1}\left((\zeta_t^2 - \zeta_t^1)dB_t^{(2)} - \frac{1}{2}|\zeta_t^2 - \zeta_t^1|^2 dt\right).$$

(4) We have

$$f(T, T_1, T_2) = -\frac{1}{T_2 - T_1}\log\left(P(T, T_2)/P(T, T_1)\right)$$
$$= f(t, T_1, T_2) - \frac{1}{T_2 - T_1}\left(\int_t^T (\zeta_s^2 - \zeta_s^1)dB_s - \frac{1}{2}(|\zeta_s^2|^2 - |\zeta_s^1|^2)ds\right)$$
$$= f(t, T_1, T_2) - \frac{1}{T_2 - T_1}\left(\int_t^T (\zeta_s^2 - \zeta_s^1)dB_s^{(2)} - \frac{1}{2}\int_t^T |\zeta_s^2 - \zeta_s^1|^2 ds\right)$$
$$= f(t, T_1, T_2) - \frac{1}{T_2 - T_1}\left(\int_t^T (\zeta_s^2 - \zeta_s^1)dB_s^{(1)} + \frac{1}{2}\int_t^T |\zeta_s^2 - \zeta_s^1|^2 ds\right).$$

Hence $f(T, T_1, T_2)$ has a Gaussian distribution given $\mathcal{F}_t$ with conditional mean

$$m_2 := f(t, T_1, T_2) + \frac{1}{2(T_2 - T_1)}\int_t^T |\zeta_s^2 - \zeta_s^1|^2 ds$$

under $\mathbb{P}_2$, resp.

$$m_1 := f(t, T_1, T_2) - \frac{1}{2(T_2 - T_1)}\int_t^T |\zeta_s^2 - \zeta_s^1|^2 ds)$$

under $\mathbb{P}_1$, and variance

$$v^2 = \frac{1}{|T_2 - T_1|^2} \int_t^T |\zeta_s^2 - \zeta_s^1|^2 ds.$$

Hence

$$(T_2 - T_1)\,\mathbb{E}\left[e^{-\int_t^{T_2} r_s ds}(f(T_1, T_1, T_2) - \kappa)^+ \Big| \mathcal{F}_t\right]$$

$$= (T_2 - T_1)P(t, T_2)\,\mathbb{E}_2\left[(f(T_1, T_1, T_2) - \kappa)^+ \Big| \mathcal{F}_t\right]$$

$$= (T_2 - T_1)P(t, T_2)\,\mathbb{E}_2\left[(m_2 + X - \kappa)^+ \Big| \mathcal{F}_t\right]$$

$$= (T_2 - T_1)P(t, T_2)\left(\frac{v}{\sqrt{2\pi}}e^{-\frac{|\kappa - m_2|^2}{2v^2}} + (m_2 - \kappa)\Phi((m_2 - \kappa)/v)\right).$$

Exercise 7.4. Bond option hedging.

(1) As in §7.3 we have

$$\mathbb{E}\left[e^{-\int_t^T r_s ds}(P(T, S) - \kappa)^+\right] = P(t, T)\,\mathbb{E}_{\tilde{\mathbb{P}}}\left[(P(T, S) - \kappa)^+\right]$$

$$= P(t, T)\,\mathbb{E}_{\tilde{\mathbb{P}}}\left[\left(\left(\frac{P(t, S)}{P(t, T)}\exp\left(\int_t^T (\sigma_s^S - \sigma_s^T)dB_s^T\right.\right.\right.$$

$$\left.\left.\left.-\frac{1}{2}\int_t^T |\sigma_s^S - \sigma_s^T|^2 ds\right) - \kappa\right)^+\right]$$

$$= P(t, T)\,\mathbb{E}[(e^X - \kappa)^+ \mid \mathcal{F}_t]$$

$$= P(t, T)e^{m_t + v^2(t, T)/2}\Phi\left(\frac{v(t, T)}{2} + \frac{1}{v(t, T)}(m_t + v^2(t, T)/2 - \log \kappa)\right)$$

$$-\kappa P(t, T)\Phi\left(-\frac{v(t, T)}{2} + \frac{1}{v(t, T)}(m_t + v^2(t, T)/2 - \log \kappa)\right),$$

with

$$m_t = \log(P(t, S)/P(t, T)) - \frac{1}{2}\int_t^T |\sigma_s^S - \sigma_s^T|^2 ds$$

and

$$v^2(t, T) = \int_t^T |\sigma_s^S - \sigma_s^T|^2 ds,$$

i.e.

$$P(t, T)\,\mathbb{E}_{\tilde{\mathbb{P}}}\left[(P(T, S) - \kappa)^+\right]$$

$$= P(t, S)\Phi\left(\frac{v(t, T)}{2} + \frac{1}{v(t, T)}\log\frac{P(t, S)}{\kappa P(t, T)}\right)$$

$$-\kappa P(t,T)\Phi\left(-\frac{v(t,T)}{2}+\frac{1}{v(t,T)}\log\frac{P(t,S)}{\kappa P(t,T)}\right)$$
$$= P(t,T)C(X_t,\kappa,v(t,T)).$$

(2) By Itô's formula we have

$$\hat{V}_T = \mathbb{E}_{\tilde{\mathbb{P}}}\left[(P(T,S)-\kappa)^+|\mathcal{F}_t\right]$$
$$= C(X_T,\kappa,0)$$
$$= C(X_0,\kappa,v(0,T)) + \int_0^t \frac{\partial C}{\partial x}(X_s,\kappa,v(s,T))dX_s$$
$$= \mathbb{E}_{\tilde{\mathbb{P}}}\left[(P(T,S)-\kappa)^+\right] + \int_0^t \frac{\partial C}{\partial x}(X_s,\kappa,v(s,T))dX_s,$$

since the processes X_t and

$$\hat{V}_t = \mathbb{E}_{\tilde{\mathbb{P}}}\left[(P(T,S)-\kappa)^+|\mathcal{F}_t\right], \qquad 0 \le t \le T,$$

are martingales under $\tilde{\mathbb{P}}$ by Proposition 7.2, and this yields (7.19), cf. also [Privault and Teng (2012)] and references therein.

(3) By numeraire invariance, cf. e.g. page 184 of [Protter (2001)], [Huang (1985)], or [Jamshidian (1996)], we have

$$dV_t = d(P(t,T)\hat{V}_t)$$
$$= \hat{V}_t dP(t,T) + P(t,T)d\hat{V}_t + dP(t,T) \cdot d\hat{V}_t$$
$$= \hat{V}_t dP(t,T) + P(t,T)\frac{\partial C}{\partial x}(X_t,\kappa,v(t,T))dX_t$$
$$\quad + \frac{\partial C}{\partial x}(X_t,\kappa,v(t,T))dP(t,T) \cdot dX_t$$
$$= \frac{\partial C}{\partial x}(X_t,\kappa,v(t,T))X_t dP(t,T)$$
$$\quad + \frac{\partial C}{\partial x}(X_t,\kappa,v(t,T))P(t,T)dX_t$$
$$\quad + \frac{\partial C}{\partial x}(X_t,\kappa,v(t,T))dP(t,T) \cdot dX_t$$
$$\quad + \left(\hat{V}_t - \frac{P(t,S)}{P(t,T)}\frac{\partial C}{\partial x}(X_t,\kappa,v(t,T))\right)dP(t,T)$$
$$= \left(\hat{V}_t - \frac{P(t,S)}{P(t,T)}\frac{\partial C}{\partial x}(X_t,\kappa,v(t,T))\right)dP(t,T)$$
$$\quad + \frac{\partial C}{\partial x}(X_t,\kappa,v(t,T))dP(t,S),$$

since

$$dP(t,S) = d(X_t P(t,T))$$
$$= X_t dP(t,T) + P(t,T)dX_t + dP(t,T) \cdot dX_t.$$

Hence, by identification with (7.18), the hedging strategy $(\xi_t^T, \xi_t^S)_{t \in [0,T]}$ of the bond option is given by

$$\xi_t^T = \hat{V}_t - \frac{P(t,S)}{P(t,T)} \frac{\partial C}{\partial x}(X_t, \kappa, v(t,T))$$

$$= C(X_t, \kappa, v(t,T)) - \frac{P(t,S)}{P(t,T)} \frac{\partial C}{\partial x}(X_t, \kappa, v(t,T)),$$

and

$$\xi_t^S = \frac{\partial C}{\partial x}(X_t, \kappa, v(t,T)), \qquad 0 \le t \le T.$$

(4) By (2.22) we have

$$\frac{\partial C}{\partial x}(x, \kappa, v) = \Phi\left(\frac{\log(x/\kappa) + \tau v^2/2}{\sqrt{\tau}v}\right),$$

and as a consequence we get

$$\xi_t^S = \Phi\left(\frac{\log(X_t/\kappa) + (T-t)v^2(t,T)/2}{v(t,T)\sqrt{T-t}}\right)$$

and

$$\xi_t^T = C(X_t, \kappa, v(t,T)) - \frac{P(t,S)}{P(t,T)} \frac{\partial C}{\partial x}(X_t, \kappa, v(t,T))$$

$$= -\kappa\Phi\left(\frac{\log(X_t/\kappa) - (T-t)v^2(t,T)/2}{v(t,T)\sqrt{T-t}}\right), \qquad 0 \le t \le T.$$

Consequently the bond option can be hedged by shortselling a bond with maturity T for the amount

$$\xi_t^T = \kappa\Phi\left(\frac{\log(X_t/\kappa) - (T-t)v^2(t,T)/2}{v(t,T)\sqrt{T-t}}\right),$$

and by buying a bond with maturity S for the amount

$$\xi_t^S = \Phi\left(\frac{\log(X_t/\kappa) + (T-t)v^2(t,T)/2}{v(t,T)\sqrt{T-t}}\right).$$

Exercise 7.5.

(1) We have

$$\mathbb{E}\left[\frac{d\mathbb{P}_T}{d\mathbb{P}}\bigg|\mathcal{F}_t\right] = e^{\sigma B_t - \sigma^2 t/2}.$$

(2) By the Girsanov Theorem 2.1, the process $\tilde{B}_t := B_t - \sigma t$ is a standard Brownian motion under $\mathbb{P}_T$.

(3) We have

$$\log P(T, S) = -\mu(S - T) + (S - T) \int_0^T \frac{\sigma}{S - s} dB_s$$

$$= -\mu(S - T) + (S - T) \int_0^t \frac{\sigma}{S - s} dB_s + (S - T) \int_t^T \frac{\sigma}{S - s} dB_s$$

$$= \frac{S - T}{S - t} \log P(t, S) + (S - T) \int_t^T \frac{\sigma}{S - s} dB_s$$

$$= \frac{S - T}{S - t} \log P(t, S) + (S - T) \int_t^T \frac{\sigma}{S - s} d\tilde{B}_s + (S - T) \int_t^T \frac{\sigma^2}{S - s} ds$$

$$= \frac{S - T}{S - t} \log P(t, S) + (S - T) \int_t^T \frac{\sigma}{S - s} d\tilde{B}_s + \sigma^2 (S - T) \log \frac{S - t}{S - T},$$

$$0 < T < S.$$

(4) We have

$$P(t, T) \, \mathbb{E}_T \left[(P(T, S) - K)^+ \big| \mathcal{F}_t \right]$$

$$= P(t, T) \, \mathbb{E}[(e^X - \kappa)^+ \mid \mathcal{F}_t]$$

$$= P(t, T) e^{m_t + v^2(t, T)/2} \Phi \left(\frac{v(t, T)}{2} + \frac{1}{v(t, T)} (m_t + v^2(t, T)/2 - \log \kappa) \right)$$

$$- \kappa P(t, T) \Phi \left(-\frac{v(t, T)}{2} + \frac{1}{v(t, T)} (m_t + v^2(t, T)/2 - \log \kappa) \right)$$

$$= P(t, T) e^{m_t + v^2(t, T)/2} \Phi \left(v(t, T) + \frac{1}{v(t, T)} (m_t - \log \kappa) \right)$$

$$- \kappa P(t, T) \Phi \left(\frac{1}{v(t, T)} (m_t - \log \kappa) \right),$$

with

$$m_t = \frac{S - T}{S - t} \log P(t, S) + \sigma^2 (S - T) \log \frac{S - t}{S - T}$$

and

$$v^2(t, T) = \sigma^2 |S - T|^2 \int_t^T \frac{1}{|S - s|^2} ds$$

$$= \sigma^2 |S - T|^2 \left(\frac{1}{S - T} - \frac{1}{S - t} \right)$$

$$= \sigma^2 (S - T) \frac{(T - t)}{(S - t)},$$

hence

$$P(t,T)\,\mathbb{E}_T\left[(P(T,S)-K)^+\big|\mathcal{F}_t\right]$$

$$= P(t,T)\,(P(t,S))^{(S-T)(S-t)}\left(\frac{S-t}{S-T}\right)^{\sigma^2(S-T)}e^{v^2(t,T)/2}$$

$$\times\Phi\left(v(t,T)+\frac{1}{v(t,T)}\log\left(\frac{(P(t,S))^{(S-T)(S-t)}}{\kappa}\left(\frac{S-t}{S-T}\right)^{\sigma^2(S-T)}\right)\right)$$

$$-\kappa P(t,T)\Phi\left(\frac{1}{v(t,T)}\log\left(\frac{(P(t,S))^{(S-T)(S-t)}}{\kappa}\left(\frac{S-t}{S-T}\right)^{\sigma^2(S-T)}\right)\right).$$

Chapter 8

Exercise 8.1. Using the decomposition

$$P(t,T)=F_1(t,X_t)F_2(t,Y_t)\exp\left(-\int_t^T\varphi(s)ds+U(t,T)\right)$$

we have, by (8.13),

$$dP(t,T)=P(t,T)r_t dt+\sigma C_1(t,T)dB_t^{(1)}+\eta C_2(t,T)dB_t^{(2)},$$

where

$$C_1(t,T)=\frac{e^{-a(T-t)}-1}{a}\quad\text{and}\quad C_2(t,T)=\frac{e^{-b(T-t)}-1}{b},\qquad 0\le t\le T.$$

Problem 8.2.

(1) We have

$$dr_t=-ar_0e^{-at}dt+\varphi'(t)dt+dX_t$$
$$=-ar_0e^{-at}dt+\theta(t)dt-a\int_0^t\theta(u)e^{-a(t-u)}dudt-aX_tdt+\sigma dB_t$$
$$=-ar_0e^{-at}dt+\theta(t)dt-a\varphi(t)dt-aX_tdt+\sigma dB_t$$
$$=(\theta(t)-ar_t)dt+\sigma dB_t.$$

(2) By standard arguments we find

$$-xF(t,x)+(\theta(t)-ax)\frac{\partial F}{\partial x}(t,x)+\frac{1}{2}\sigma^2\frac{\partial^2 F}{\partial x^2}(t,x)+\frac{\partial F}{\partial t}(t,x)=0,\quad(14.5)$$

under the terminal condition $F(T,x)=1$, $x\in\mathbb{R}$.

(3) We have

$$
P(t,T) = \mathbb{E}\left[e^{-\int_t^T r_s ds}\Big|\mathcal{F}_t\right]
$$

$$
= \mathbb{E}\left[e^{-\int_t^T (r_0 e^{-as}+\varphi(s)+X_s)ds}\Big|\mathcal{F}_t\right]
$$

$$
= e^{-\int_t^T (r_0 e^{-as}+\varphi(s))ds}\ \mathbb{E}\left[e^{-\int_t^T X_s ds}\Big|\mathcal{F}_t\right]
$$

$$
= e^{-\int_t^T (r_0 e^{-as}+\varphi(s))ds}\ \exp\left(-\mathbb{E}\left[\int_t^T X_s ds\Big|\mathcal{F}_t\right] + \frac{1}{2}\operatorname{Var}\left[\int_t^T X_s ds\Big|\mathcal{F}_t\right]\right)
$$

$$
= e^{-\int_t^T (r_0 e^{-as}+\varphi(s))ds}\ e^{-\frac{X_t}{a}(1-e^{-a(T-t)})+\frac{1}{2}\frac{\sigma^2}{a^2}\int_t^T |e^{-a(T-s)}-1|^2 ds}
$$

$$
= e^{A(t,T)+X_t C(t,T)},
$$

where

$$
A(t,T) = -\int_t^T (r_0 e^{-as}+\varphi(s))ds - \frac{X_t}{a}(1-e^{-a(T-t)}),
$$

and

$$
C(t,T) = \frac{\sigma^2}{2a^2}\int_t^T |e^{-a(T-s)}-1|^2 ds.
$$

(4) We have

$$
f(t,T) = -\frac{\partial \log P(t,T)}{\partial T} = r_0 e^{-aT} + \varphi(T) + X_t e^{-a(T-t)}
$$

$$
- \frac{\sigma^2}{2a^2}|1-e^{-a(T-t)}|^2,
$$

$0 \le t \le T.$

(5) We have

$$
d_t f(t,T) = aX_t e^{-a(T-t)}dt + e^{-a(T-t)}dX_t + \frac{\sigma^2}{a}e^{-a(T-t)}(1-e^{-a(T-t)})dt
$$

$$
= aX_t e^{-a(T-t)}dt + e^{-a(T-t)}(-aX_t + \sigma dB_t) + \frac{\sigma^2}{a}e^{-a(T-t)}(1-e^{-a(T-t)})dt
$$

$$
= e^{-a(T-t)}\sigma dB_t + \frac{\sigma^2}{a}e^{-a(T-t)}(1-e^{-a(T-t)})dt.
$$

(6) We have

$$
\sigma^2 e^{-a(T-t)}\int_t^T e^{-a(T-s)} = \frac{\sigma^2}{a}e^{-a(T-t)}(1-e^{-a(T-t)}).
$$

(7) Since $t=0$, it suffices to let

$$
\varphi(T) = -r_0 e^{-aT} + f^M(0,T) + \frac{\sigma^2}{2a^2}|1-e^{-aT}|^2, \qquad T>0,
$$

to obtain $f(0,T) = f^M(0,T)$, $T>0$.

(8) Differentiating the relation

$$\varphi(T) = \int_0^T \theta(t)e^{-a(T-t)}dt = -r_0e^{-aT} + f^M(0,T) + \frac{\sigma^2}{2a^2}|1 - e^{-aT}|^2,$$

$T > 0$, we get

$$\theta(T) - a\varphi(T) = ar_0e^{-aT} + \frac{\partial f^M}{\partial t}(0,T) + \frac{\sigma^2}{a}e^{-aT}(1 - e^{-aT}), \qquad T > 0,$$

hence

$$\begin{aligned}
\theta(t) &= a\varphi(t) + ar_0e^{-at} + \frac{\partial f^M}{\partial t}(0,t) + \frac{\sigma^2}{a}e^{-at}(1 - e^{-at}) \\
&= af(0,t) + \frac{\partial f^M}{\partial t}(0,t) + \frac{\sigma^2}{2a}(1 - e^{-2at}) \\
&= af^M(0,t) + \frac{\partial f^M}{\partial t}(0,t) + \frac{\sigma^2}{2a}(1 - e^{-2at}), \qquad t > 0.
\end{aligned}$$

(9) From the Itô formula, the PDE (14.5) and the martingale property of $t \longmapsto e^{-\int_0^t r_s ds}P(t,T)$ we have

$$\begin{aligned}
d\left(e^{-\int_0^t r_s ds}P(t,T)\right) &= d\left(e^{-\int_0^t r_s ds}F(t,r_t)\right) \\
&= \sigma e^{-\int_0^t r_s ds}\frac{\partial F}{\partial x}(t,r_t)dB_t \\
&= \sigma e^{-\int_0^t r_s ds}P(t,T)\frac{\partial \log F}{\partial x}(t,r_t)dB_t,
\end{aligned}$$

hence

$$\zeta_t = \sigma \frac{\partial \log F}{\partial x}(t,r_t) = \sigma C(t,T),$$

and we have

$$\begin{aligned}
dP(t,T) &= e^{\int_0^t r_s ds}d\left(e^{-\int_0^t r_s ds}P(t,T)\right) + r_tP(t,T)dt \\
&= r_tP(t,T)dt + \zeta_tP(t,T)dB_t.
\end{aligned}$$

(10) From Question 9 we have

$$e^{\int_0^t r_s ds}P(t,T) = P(0,T)e^{\int_0^t \zeta_s dB_s + \frac{1}{2}\int_0^t \zeta_s^2 ds},$$

hence

$$\frac{d\tilde{\mathbb{P}}}{d\mathbb{P}} = \mathbb{E}\left[\frac{d\tilde{\mathbb{P}}}{d\mathbb{P}}\bigg|\mathcal{F}_T\right] = \frac{1}{P(0,T)}e^{-\int_0^T r_s ds} = e^{\int_0^t \zeta_s dB_s - \frac{1}{2}\int_0^t \zeta_s^2 ds}.$$

(11) We have

$$dr_t = (\theta(t) - ar_t) + \sigma dB_t = (\theta(t) - ar_t) + \sigma(\sigma C(t,T)dt + d\hat{B}_t),$$

where $\hat{B}_t$ is a standard Brownian motion under $\tilde{\mathbb{P}}$.

(12) We have
$$d\frac{P(t,S)}{P(t,T)} = \frac{P(t,S)}{P(t,T)}(\zeta_t^S - \zeta_t^T)(dB_t - \zeta_t^T dt) = \frac{P(t,S)}{P(t,T)}(\zeta_t^S - \zeta_t^T)d\hat{B}_t,$$
$$0 \le t \le T.$$

(13) We have
$$\mathbb{E}_{\tilde{\mathbb{P}}}\left[P(T,S)\big|\mathcal{F}_t\right] = \mathbb{E}_{\tilde{\mathbb{P}}}\left[\frac{P(T,S)}{P(T,T)}\Big|\mathcal{F}_t\right] = \frac{P(t,S)}{P(t,T)}, \qquad 0 \le t \le T \le S,$$

hence
$$\frac{P(t,S)}{P(t,T)} = \mathbb{E}_{\tilde{\mathbb{P}}}\left[P(T,S)\big|\mathcal{F}_T\right]$$
$$= \mathbb{E}_{\tilde{\mathbb{P}}}\left[e^{A(T,S)+X_T C(T,S)}\big|\mathcal{F}_T\right]$$
$$= e^{A(T,S)+C(T,S)\mathbb{E}[X_T|\mathcal{F}_t]+\frac{1}{2}|C(T,S)|^2\,\mathrm{Var}[X_T|\mathcal{F}_t]},$$

hence
$$A(T,S)+C(T,S)\,\mathbb{E}[X_T\mid\mathcal{F}_t]+\frac{1}{2}|C(T,S)|^2\,\mathrm{Var}[X_T\mid\mathcal{F}_t] = \log\frac{P(t,S)}{P(t,T)}.$$

(14) We have, using the call-put duality
$$P(t,T)\,\mathbb{E}_{\tilde{\mathbb{P}}}\left[(K - P(T,S))^+\big|\mathcal{F}_t\right]$$
$$= P(t,T)\,\mathbb{E}_{\tilde{\mathbb{P}}}\left[(K - P(T,S))\big|\mathcal{F}_t\right] + P(t,T)\,\mathbb{E}_{\tilde{\mathbb{P}}}\left[(P(T,S) - K)^+\big|\mathcal{F}_t\right]$$
$$= KP(t,T) - P(t,T)\,\mathbb{E}_{\tilde{\mathbb{P}}}\left[P(T,S)\big|\mathcal{F}_t\right] + P(t,T)\,\mathbb{E}_{\tilde{\mathbb{P}}}\left[(P(T,S) - K)^+\big|\mathcal{F}_t\right]$$
$$= KP(t,T) - P(t,S) + P(t,T)\,\mathbb{E}_{\tilde{\mathbb{P}}}\left[(e^X - K)^+\big|\mathcal{F}_t\right],$$

where X is a centered Gaussian random variable with mean
$$m_t = A(T,S) + C(T,S)\,\mathbb{E}[X_T\mid\mathcal{F}_t]$$

and variance
$$v^2(t,T) = |C(T,S)|^2\,\mathrm{Var}[X_T\mid\mathcal{F}_t]$$

given $\mathcal{F}_t$, hence
$$P(t,T)\,\mathbb{E}_{\tilde{\mathbb{P}}}\left[(K - P(T,S))^+\big|\mathcal{F}_t\right]$$
$$= KP(t,T) - P(t,S) + P(t,T)\Phi\left(\frac{v(t,T)}{2}\right.$$
$$\left. +\frac{1}{v(t,T)}(m_t + v^2(t,T)/2 - \log K)\right)$$
$$-P(t,T)\Phi\left(-\frac{v(t,T)}{2} + \frac{1}{v(t,T)}(m_t + v^2(t,T)/2 - \log K)\right)$$
$$= KP(t,T) - P(t,S) + P(t,T)\Phi\left(\frac{v(t,T)}{2} + \frac{1}{v(t,T)}\log\frac{P(t,S)}{KP(t,T)}\right)$$
$$-P(t,T)\Phi\left(-\frac{v(t,T)}{2} + \frac{1}{v(t,T)}\log\frac{P(t,S)}{KP(t,T)}\right).$$

Chapter 9

Problem 9.1. Defaultable bonds.

(1) Use the fact that $(r_t, \lambda_t)_{t \in [0,T]}$ is a Markov process.
(2) Use the tower property for the conditional expectation under $\mathcal{F}_t$.
(3) We have

$$d \left(e^{- \int_0^t (r_s + \lambda_s) ds} P(t,T) \right)$$

$$= -(r_t + \lambda_t) e^{- \int_0^t (r_s + \lambda_s) ds} P(t,T) dt + e^{- \int_0^t (r_s + \lambda_s) ds} dP(t,T)$$

$$= -(r_t + \lambda_t) e^{- \int_0^t (r_s + \lambda_s) ds} P(t,T) dt + e^{- \int_0^t (r_s + \lambda_s) ds} dF(t, r_t, \lambda_t)$$

$$= -(r_t + \lambda_t) e^{- \int_0^t (r_s + \lambda_s) ds} P(t,T) dt + e^{- \int_0^t (r_s + \lambda_s) ds} \frac{\partial F}{\partial x}(t, r_t, \lambda_t) dr_t$$

$$+ e^{- \int_0^t (r_s + \lambda_s) ds} \frac{\partial F}{\partial y}(t, r_t, \lambda_t) d\lambda_t + \frac{1}{2} e^{- \int_0^t (r_s + \lambda_s) ds} \frac{\partial^2 F}{\partial x^2}(t, r_t, \lambda_t) \sigma_1^2(t, r_t) dt$$

$$+ \frac{1}{2} e^{- \int_0^t (r_s + \lambda_s) ds} \frac{\partial^2 F}{\partial y^2}(t, r_t, \lambda_t) \sigma_2^2(t, \lambda_t) dt$$

$$+ e^{- \int_0^t (r_s + \lambda_s) ds} \rho \frac{\partial^2 F}{\partial x \partial y}(t, r_t, \lambda_t) \sigma_1(t, r_t) \sigma_2(t, \lambda_t) dt$$

$$+ e^{- \int_0^t (r_s + \lambda_s) ds} \frac{\partial F}{\partial t}(t, r_t, \lambda_t) dt$$

$$= e^{- \int_0^t (r_s + \lambda_s) ds} \frac{\partial F}{\partial x}(t, r_t, \lambda_t) \sigma_1(t, r_t) dB_t^1$$

$$+ e^{- \int_0^t (r_s + \lambda_s) ds} \frac{\partial F}{\partial y}(t, r_t, \lambda_t) \sigma_2(t, \lambda_t) dB_t^2$$

$$e^{- \int_0^t (r_s + \lambda_s) ds} \left(-(r_t + \lambda_t) P(t,T) + \frac{\partial F}{\partial x}(t, r_t, \lambda_t) \mu_1(t, r_t) \right.$$

$$+ \frac{\partial F}{\partial y}(t, r_t, \lambda_t) \mu_2(t, \lambda_t)$$

$$+ \frac{1}{2} \frac{\partial^2 F}{\partial x^2}(t, r_t, \lambda_t) \sigma_1^2(t, r_t) + \frac{1}{2} \frac{\partial^2 F}{\partial y^2}(t, r_t, \lambda_t) \sigma_2^2(t, \lambda_t)$$

$$\left. + \rho \frac{\partial^2 F}{\partial x \partial y}(t, r_t, \lambda_t) \sigma_1(t, r_t) \sigma_2(t, \lambda_t) + \frac{\partial F}{\partial t}(t, r_t, \lambda_t) \right) dt,$$

hence the bond pricing PDE is

$$-(x+y) F(t,x,y) + \mu_1(t,x) \frac{\partial F}{\partial x}(t,x,y)$$

$$+ \mu_2(t,y) \frac{\partial F}{\partial y}(t,x,y) + \frac{1}{2} \sigma_1^2(t,x) \frac{\partial^2 F}{\partial x^2}(t,x,y)$$

$$+ \frac{1}{2} \sigma_2^2(t,y) \frac{\partial^2 F}{\partial y^2}(t,x,y) + \rho \sigma_1(t,x) \sigma_2(t,y) \frac{\partial^2 F}{\partial x \partial y}(t,x,y) + \frac{\partial F}{\partial t}(t, r_t, \lambda_t) = 0.$$

(4) We have

$$\int_0^t r_s ds = \frac{1}{a}\left(\sigma B_t^1 - r_t\right)$$

$$= \frac{\sigma}{a}\left(B_t^1 - \int_0^t e^{-a(t-s)}dB_s^1\right)$$

$$= \frac{\sigma}{a}\int_0^t (1 - e^{-a(t-s)})dB_s^1,$$

hence

$$\int_t^T r_s ds = \int_0^T r_s ds - \int_0^t r_s ds$$

$$= \frac{\sigma}{a}\int_0^T (1 - e^{-a(T-s)})dB_s^1 - \frac{\sigma}{a}\int_0^t (1 - e^{-a(t-s)})dB_s^1$$

$$= -\frac{\sigma}{a}\left(\int_0^t (e^{-a(T-s)} - e^{-a(t-s)})dB_s^1 + \int_t^T (e^{-a(T-s)} - 1)dB_s^1\right)$$

$$= -\frac{\sigma}{a}(e^{-a(T-t)} - 1)\int_0^t e^{-a(t-s)}dB_s^1 - \frac{\sigma}{a}\int_t^T (e^{-a(T-s)} - 1)dB_s^1$$

$$= \frac{1}{a}(1 - e^{-a(T-t)})r_t - \frac{\sigma}{a}\int_t^T (e^{-a(T-s)} - 1)dB_s^1.$$

The answer for λ_t is similar.

(5) As a consequence of the previous question we have

$$\mathbb{E}\left[\int_t^T r_s ds + \int_t^T \lambda_s ds \,\Big|\, \mathcal{F}_t\right] = C(a,t,T)r_t + C(b,t,T)\lambda_t,$$

and

$$\mathrm{Var}\left[\int_t^T r_s ds + \int_t^T \lambda_s ds \,\Big|\, \mathcal{F}_t\right]$$

$$= \mathrm{Var}\left[\int_t^T r_s ds \,\Big|\, \mathcal{F}_t\right] + \mathrm{Var}\left[\int_t^T \lambda_s ds \,\Big|\, \mathcal{F}_t\right]$$

$$+ 2\,\mathrm{Cov}\left(\int_t^T X_s ds, \int_t^T Y_s ds \,\Big|\, \mathcal{F}_t\right)$$

$$= \frac{\sigma^2}{a^2}\int_t^T |e^{-a(T-s)} - 1|^2 ds$$

$$+ 2\rho\frac{\sigma\eta}{ab}\int_t^T (e^{-a(T-s)} - 1)(e^{-b(T-s)} - 1)ds$$

$$+\frac{\eta^2}{b^2}\int_t^T |e^{-b(T-s)}-1|^2 ds$$

$$=\sigma^2\int_t^T C^2(a,s,T)ds + 2\rho\sigma\eta\int_t^T C(a,s,T)C(b,s,T)ds$$

$$+\eta^2\int_t^T C^2(b,sT)ds,$$

from the Itô isometry.

(6) We have

$$P(t,T) = \mathbf{1}_{\{\tau>t\}}\, \mathbb{E}\left[\exp\left(-\int_t^T r_s ds - \int_t^T \lambda_s ds\right)\Big|\mathcal{F}_t\right]$$

$$= \mathbf{1}_{\{\tau>t\}}\exp\left(-\mathbb{E}\left[\int_t^T r_s ds\Big|\mathcal{F}_t\right] - \mathbb{E}\left[\int_t^T \lambda_s ds\Big|\mathcal{F}_t\right]\right)$$

$$\times \exp\left(\frac{1}{2}\operatorname{Var}\left[\int_t^T r_s ds + \int_t^T \lambda_s ds\Big|\mathcal{F}_t\right]\right)$$

$$= \mathbf{1}_{\{\tau>t\}}\exp\left(-C(a,t,T)r_t - C(b,t,T)\lambda_t\right)$$

$$\times \exp\left(\frac{\sigma^2}{2}\int_t^T C^2(a,s,T)ds + \frac{\eta^2}{2}\int_t^T C^2(b,s,T)e^{-b(T-s)}ds\right)$$

$$\times \exp\left(\rho\sigma\eta\int_t^T C(a,s,T)C(b,s,T)ds\right).$$

(7) This is a direct consequence of the answers to Questions 3 and 6.

(8) The above analysis shows that

$$\mathbb{P}(\tau>T\mid\mathcal{G}_t) = \mathbf{1}_{\{\tau>t\}}\,\mathbb{E}\left[\exp\left(-\int_t^T \lambda_s ds\right)\Big|\mathcal{F}_t\right]$$

$$= \mathbf{1}_{\{\tau>t\}}\exp\left(-C(b,t,T)\lambda_t + \frac{\eta^2}{2}\int_t^T C^2(b,s,T)ds\right),$$

for $a=0$ and

$$\mathbb{E}\left[\exp\left(-\int_t^T r_s ds\right)\Big|\mathcal{F}_t\right] = \exp\left(-C(a,t,T)r_t + \frac{\sigma^2}{2}\int_t^T C^2(a,s,T)ds\right),$$

for $b=0$, and this implies

$$U(t,T) = \exp\left(\rho\sigma\eta\int_t^T C(a,s,T)C(b,s,T)ds\right)$$

$$= \exp\left(\rho\frac{\sigma\eta}{ab}\left(T - t - C(a,t,T) - C(b,t,T) + C(a+b,t,T)\right)\right).$$

(9) We have

$$f(t,T) = -\mathbf{1}_{\{\tau>t\}}\frac{\partial \log P(t,T)}{\partial T}$$

$$= \mathbf{1}_{\{\tau>t\}}\left(r_t e^{-a(T-t)} - \frac{\sigma^2}{2}C^2(a,t,T) + \lambda_t e^{-b(T-t)} - \frac{\eta^2}{2}C^2(b,t,T)\right)$$

$$-\mathbf{1}_{\{\tau>t\}}\rho\sigma\eta C(a,t,T)C(b,t,T).$$

(10) We use the relation

$$\mathbb{P}(\tau > T \mid \mathcal{G}_t) = \mathbf{1}_{\{\tau>t\}}\,\mathbb{E}\left[\exp\left(-\int_t^T \lambda_s ds\right)\Big|\mathcal{F}_t\right]$$

$$= \mathbf{1}_{\{\tau>t\}}\exp\left(-C(b,t,T)\lambda_t + \frac{\eta^2}{2}\int_t^T C^2(b,s,T)ds\right)$$

$$= \mathbf{1}_{\{\tau>t\}}e^{-\int_t^T f_2(t,u)du},$$

where $f_2(t,T)$ is the Vasicek forward rate corresponding to λ_t, i.e.

$$f_2(t,u) = \lambda_t e^{-b(u-t)} - \frac{\eta^2}{2}C^2(b,t,u).$$

(11) In this case we have $\rho = 0$ and

$$P(t,T) = \mathbb{P}(\tau > T \mid \mathcal{G}_t)\,\mathbb{E}\left[\exp\left(-\int_t^T r_s ds\right)\Big|\mathcal{F}_t\right],$$

since $U(t,T) = 0$.

Chapter 10

Exercise 10.1.

(1) We have

$$L(t,T_1,T_2) = L(0,T_1,T_2)e^{\int_0^t \gamma_1(s)dB_s^{(2)} - \frac{1}{2}\int_0^t |\gamma_1(s)|^2 ds}, \qquad 0 \le t \le T_1,$$

and $L(t,T_2,T_3) = b$, hence $P(t,T_2)/P(t,T_3) = 1 + \delta b$ and $\mathbb{P}_2 = \mathbb{P}_3$ up to time T_2, while $\mathbb{P}_2 = \mathbb{P}_{1,2}$ up to time T_1 by (10.22).

(2) We have

$$\mathbb{E}\left[e^{-\int_t^{T_2} r_s ds}(L(T_1,T_1,T_2) - \kappa)^+ \Big|\mathcal{F}_t\right]$$

$$= P(t,T_2)\,\mathbb{E}_2\left[(L(T_1,T_1,T_2) - \kappa)^+ \mid \mathcal{F}_t\right]$$

$$= P(t, T_2) \, \mathbb{E}_2 \left[\left(L(t, T_1, T_2) e^{\int_t^{T_1} \gamma_1(s) dB_s^{(2)} - \frac{1}{2} \int_t^{T_1} |\gamma_1(s)|^2 ds} - \kappa \right)^+ \Big| \mathcal{F}_t \right]$$

$$= P(t, T_2) \mathrm{Bl}(\kappa, L(t, T_1, T_2), \sigma_1(t), 0, T_1 - t),$$

where

$$\sigma_1^2(t) = \frac{1}{T_1 - t} \int_t^{T_1} |\gamma_1(s)|^2 ds,$$

and

$$\mathbb{E} \left[e^{-\int_t^{T_3} r_s ds} (L(T_2, T_2, T_3) - \kappa)^+ \Big| \mathcal{F}_t \right] = P(t, T_3) \, \mathbb{E}_3 \left[(b - \kappa)^+ \mid \mathcal{F}_t \right]$$

$$= P(t, T_3)(b - \kappa)^+.$$

(3) We have

$$\frac{P(t, T_1)}{P(t, T_1, T_3)} = \frac{P(t, T_1)}{\delta P(t, T_2) + \delta P(t, T_3)}$$

$$= \frac{P(t, T_1)}{\delta P(t, T_2)} \frac{1}{1 + P(t, T_3)/P(t, T_2)}$$

$$= \frac{1 + \delta b}{\delta(\delta b + 2)} (1 + \delta L(t, T_1, T_2)), \qquad 0 \leq t \leq T_1,$$

and

$$\frac{P(t, T_3)}{P(t, T_1, T_3)} = \frac{P(t, T_3)}{P(t, T_2) + P(t, T_3)}$$

$$= \frac{1}{1 + P(t, T_2)/P(t, T_3)}$$

$$= \frac{1}{\delta} \frac{1}{\delta b + 2}, \qquad 0 \leq t \leq T_2. \tag{14.6}$$

(4) We have

$$S(t, T_1, T_3) = \frac{P(t, T_1)}{P(t, T_1, T_3)} - \frac{P(t, T_3)}{P(t, T_1, T_3)}$$

$$= \frac{1 + \delta b}{\delta(2 + \delta b)} (1 + \delta L(t, T_1, T_2)) - \frac{1}{\delta(2 + \delta b)}$$

$$= \frac{1}{2 + \delta b} (b + (1 + \delta b) L(t, T_1, T_2)), \qquad 0 \leq t \leq T_2,$$

and

$$dS(t, T_1, T_3) = \frac{1 + \delta b}{2 + \delta b} L(t, T_1, T_2) \gamma_1(t) dB_t^{(2)}$$

$$= \left(S(t, T_1, T_3) - \frac{b}{2 + \delta b} \right) \gamma_1(t) dB_t^{(2)}$$

$$= S(t, T_1, T_3) \sigma_{1,3}(t) dB_t^{(2)},$$

$0 \le t \le T_2$, with

$$
\begin{aligned}
\sigma_{1,3}(t) &= \left(1 - \frac{b}{S(t,T_1,T_3)(2+\delta b)}\right)\gamma_1(t) \\
&= \left(1 - \frac{b}{b+(1+\delta b)L(t,T_1,T_2)}\right)\gamma_1(t) \\
&= \frac{(1+\delta b)L(t,T_1,T_2)}{b+(1+\delta b)L(t,T_1,T_2)}\gamma_1(t) \\
&= \frac{(1+\delta b)L(t,T_1,T_2)}{(2+\delta b)S(t,T_1,T_2)}\gamma_1(t).
\end{aligned}
$$

Exercise 10.2.

(1) We have

$$
L(t,T_1,T_2) = L(0,T_1,T_2)e^{\gamma B_t^{(2)} - \gamma^2 t/2}, \qquad 0 \le t \le T_1.
$$

(2) We have

$$
\begin{aligned}
&P(t,T_2)\,\mathbb{E}_2\left[(L(T_1,T_1,T_2) - \kappa)^+ \mid \mathcal{F}_t\right] \\
&= P(t,T_2)\mathrm{Bl}(\kappa, L(t,T_1,T_2), \gamma, 0, T_1 - t), \qquad 0 \le t \le T_1.
\end{aligned}
$$

Exercise 10.3.

(1) We have

$$
\begin{aligned}
L(T,T_1,T_2) &= S(T,T_1,T_2) \\
&= \frac{1}{T_2 - T_1}\left(\frac{P(T,T_1)}{P(T,T_2)} - 1\right) \\
&= \frac{1}{T_2 - T_1}\left(\frac{P(t,T_1)}{P(t,T_2)}\exp\left(\int_t^T(\zeta_s^1 - \zeta_s^2)dB_s - \frac{1}{2}\int_t^T(|\zeta_s^1|^2 - |\zeta_s^2|^2)ds\right) - 1\right) \\
&= \frac{1}{T_2 - T_1}\left(\frac{P(t,T_1)}{P(t,T_2)}\exp\left(\int_t^T(\zeta_s^1 - \zeta_s^2)dB_s^{(2)} - \frac{1}{2}\int_t^T|\zeta_s^1 - \zeta_s^2|^2 ds\right) - 1\right) \\
&= \frac{1}{T_2 - T_1}\left(\frac{P(t,T_1)}{P(t,T_2)}\exp\left(\int_t^T(\zeta_s^1 - \zeta_s^2)dB_s^{(1)} + \frac{1}{2}\int_t^T|\zeta_s^1 - \zeta_s^2|^2 ds\right) - 1\right),
\end{aligned}
$$

and by Itô calculus,

$$
\begin{aligned}
dS(t,T_1,T_2) &= \frac{1}{T_2 - T_1}d\left(\frac{P(t,T_1)}{P(t,T_2)}\right) \\
&= \frac{1}{T_2 - T_1}\frac{P(t,T_1)}{P(t,T_2)}\left((\zeta_t^1 - \zeta_t^2)dB_t + \frac{1}{2}|\zeta_t^1 - \zeta_t^2|^2 dt - \frac{1}{2}(|\zeta_t^1|^2 - |\zeta_t^2|^2)dt\right)
\end{aligned}
$$

$$= \left(\frac{1}{T_2 - T_1} + S(t, T_1, T_2)\right) \left((\zeta_t^1 - \zeta_t^2)dB_t + \zeta_t^2(\zeta_t^2 - \zeta_t^1)dt)dt\right)$$

$$= \left(\frac{1}{T_2 - T_1} + S(t, T_1, T_2)\right) \left((\zeta_t^1 - \zeta_t^2)dB_t^{(1)} + (|\zeta_t^2|^2 - |\zeta_t^1|^2)dt\right)$$

$$= \left(\frac{1}{T_2 - T_1} + S(t, T_1, T_2)\right) (\zeta_t^1 - \zeta_t^2)dB_t^{(2)}, \qquad 0 \le t \le T_1,$$

hence $\frac{1}{T_2-T_1} + S(t, T_1, T_2)$ is a geometric Brownian motion, with

$$\frac{1}{T_2 - T_1} + S(T, T_1, T_2)$$

$$= \left(\frac{1}{T_2 - T_1} + S(t, T_1, T_2)\right) \exp\left(\int_t^T (\zeta_s^1 - \zeta_s^2)dB_s^{(2)} - \frac{1}{2}\int_t^T |\zeta_s^1 - \zeta_s^2|^2 ds\right),$$

$0 \le t \le T \le T_1.$

(2) We have

$$(T_2 - T_1)\, \mathbb{E}\left[e^{-\int_t^{T_2} r_s ds}(L(T_1, T_1, T_2) - \kappa)^+ \Big| \mathcal{F}_t\right]$$

$$= (T_2 - T_1)\, \mathbb{E}\left[e^{-\int_t^{T_1} r_s ds}P(T_1, T_2)(L(T_1, T_1, T_2) - \kappa)^+ \Big| \mathcal{F}_t\right]$$

$$= P(t, T_1, T_2)\, \mathbb{E}_{1,2}\left[(S(T_1, T_1, T_2) - \kappa)^+ \Big| \mathcal{F}_t\right].$$

Since $(B_t^2)_{t \in [0, T_1]}$ is a standard Brownian motion until time T_1 under $\mathbb{P}_{1,2}$,

$$L(T, T_1, T_2) = S(T, T_1, T_2)$$

$$= -\frac{1}{T_2 - T_1}$$

$$+ \left(\frac{1}{T_2 - T_1} + S(t, T_1, T_2)\right) \exp\left(\int_t^T (\zeta_s^1 - \zeta_s^2)dB_s^{(2)} - \frac{1}{2}\int_t^T |\zeta_s^1 - \zeta_s^2|^2 ds\right),$$

has same law as

$$\frac{1}{T_2 - T_1}\left(\frac{P(t, T_1)}{P(t, T_2)}e^{X - \frac{1}{2}\operatorname{Var}[X]} - 1\right),$$

where X is a centered Gaussian random variable under $\mathbb{P}_2$, with variance

$$\int_t^{T_1} |\zeta_s^1 - \zeta_s^2|^2 ds$$

given $\mathcal{F}_t$. Hence

$$(T_2 - T_1)\, \mathbb{E}\left[e^{-\int_t^{T_2} r_s ds}(L(T_1, T_1, T_2) - \kappa)^+ \Big| \mathcal{F}_t\right]$$

$$= P(t, T_1, T_2)$$

$$\times \operatorname{Bl}\left(\frac{1}{T_2 - T_1} + S(t, T_1, T_2), \frac{\int_t^{T_1} |\zeta_s^1 - \zeta_s^2|^2 ds}{T_1 - t}, \kappa + \frac{1}{T_2 - T_1}, T_1 - t\right).$$

Exercise 10.4. Swaption hedging.

(1) We have

$$S(T_i, T_i, T_j) = S(t, T_i, T_j) \exp\left(\int_t^{T_i} \sigma_{i,j}(s) dB_s^{i,j} - \frac{1}{2}\int_t^{T_i} |\sigma_{i,j}|^2(s) ds\right),$$

and

$$P(t, T_i, T_j)\, \mathbb{E}_{i,j}\left[\left(S(T_i, T_i, T_j) - \kappa\right)^+ \Big| \mathcal{F}_t\right]$$

$$= P(t, T_i, T_j)$$

$$\times \mathbb{E}_{i,j}\left[\left(S(t, T_i, T_j)\exp\left(\int_t^{T_i}\sigma_{i,j}(s)dB_s^{i,j} - \frac{1}{2}\int_t^{T_i}|\sigma_{i,j}|^2(s)ds\right) - \kappa\right)^+ \Big| \mathcal{F}_t\right]$$

$$= P(t, T_i, T_j)\mathrm{Bl}(\kappa, v(t, T_i)/\sqrt{T_i - t}, 0, T_i - t)$$

$$= P(t, T_i, T_j)C(S(t, T_i, T_j), \kappa, v(t, T_i)),$$

by Lemma 2.3, where

$$v^2(t, T_i) = \int_t^{T_i}|\sigma_{i,j}|^2(s)ds,$$

and

$$C(x, \kappa, v) = x\Phi\left(\frac{\log(x/K)}{v} + \frac{v}{2}\right) - \kappa\Phi\left(\frac{\log(x/K)}{v} - \frac{v}{2}\right).$$

(2) We apply the Itô formula to the forward portfolio price $\hat{V}_t = C(S_t, \kappa, v(t, T_i))$, and use the fact that both $\hat{V}_t$ and $(S(t, T_i, T_j))_{t\in[0,T_i]}$ are martingales under $\mathbb{P}_{i,j}$.

(3) By numeraire invariance, letting $S_t = S(t, T_i, T_j)$ we have, as in Exercise 7.4-(2),

$$dV_t = d(P(t, T_i, T_j)\hat{V}_t)$$

$$= \hat{V}_t dP(t, T_i, T_j) + P(t, T_i, T_j)d\hat{V}_t + dP(t, T_i, T_j) \cdot d\hat{V}_t$$

$$= \hat{V}_t dP(t, T_i, T_j) + P(t, T_i, T_j)\frac{\partial C}{\partial x}(S_t, \kappa, v(t, T_i))dS_t$$

$$+ \frac{\partial C}{\partial x}(S_t, \kappa, v(t, T_i))dP(t, T_i, T_j) \cdot dS_t$$

$$= \frac{\partial C}{\partial x}(S_t, \kappa, v(t, T_i))S_t dP(t, T_i, T_j) + P(t, T_i, T_j)\frac{\partial C}{\partial x}(S_t, \kappa, v(t, T_i))dS_t$$

$$+ \frac{\partial C}{\partial x}(S_t, \kappa, v(t, T_i))dP(t, T_i, T_j) \cdot dS_t$$

$$+ \left(\hat{V}_t - S_t\frac{\partial C}{\partial x}(S_t, \kappa, v(t, T_i))\right)dP(t, T_i, T_j)$$

$$= \left(\hat{V}_t - S_t \frac{\partial C}{\partial x}(S_t, \kappa, v(t,T_i)) \right) dP(t,T_i,T_j)$$
$$+ \frac{\partial C}{\partial x}(S_t, \kappa, v(t,T_i)) d(P(t,T_i) - P(t,T_j)),$$

since

$$d(P(t,T_i) - P(t,T_j)) = d(S_t P(t,T_i,T_j))$$
$$= S_t dP(t,T_i,T_j) + P(t,T_i,T_j) dS_t + dP(t,T_i,T_j) \cdot dS_t,$$

cf. also [Privault and Teng (2012)] and references therein.

(4) By (2.22) we have

$$dV_t = \Phi \left(\frac{\log(S_t/K)}{v(t,T_i)} + \frac{v(t,T_i)}{2} \right) d(P(t,T_i) - P(t,T_j)) \quad (14.7)$$
$$- \kappa \Phi \left(\frac{\log(S_t/K)}{v(t,T_i)} - \frac{v(t,T_i)}{2} \right) dP(t,T_i,T_j).$$

Consequently, by comparing (10.27) and (14.7) the hedging strategy of the swaption is given by

$$\xi_t^i = \Phi \left(\frac{\log(S_t/K)}{v(t,T_i)} + \frac{v(t,T_i)}{2} \right),$$

$$\xi_t^j = -\Phi \left(\frac{\log(S_t/K)}{v(t,T_i)} + \frac{v(t,T_i)}{2} \right) - \kappa(T_{j+1} - T_j) \Phi \left(\frac{\log(S_t/K)}{v(t,T_i)} - \frac{v(t,T_i)}{2} \right),$$

and

$$\xi_t^k = -\kappa(T_{k+1} - T_k) \Phi \left(\frac{\log(S_t/K)}{v(t,T_i)} - \frac{v(t,T_i)}{2} \right), \qquad i+1 \le k \le j-1.$$

Chapter 11

Exercise 11.1. From Relations (10.18) and (14.6) above we have $\mathbb{P}_3 = \mathbb{P}_{1,3}$ up to time T_1, hence $(B^3)_{t \in [0,T_1]}$ is a standard Brownian motion under $\mathbb{P}_{1,3}$ and

$$P(t,T_1,T_3) \mathbb{E}_{1,3} \left[(S(T_1,T_1,T_3) - \kappa)^+ \mid \mathcal{F}_t \right]$$
$$= \text{Bl}(\kappa, S(t,T_1,T_2), \tilde{\sigma}_{1,3}(t), 0, T_1 - t), \qquad 0 \le t \le T_1,$$

where $\tilde{\sigma}_{1,3}(t)$ is the approximate volatility obtained by freezing the random component of $\sigma_{1,3}(s)$ at time t, i.e.

$$
\begin{aligned}
|\tilde{\sigma}_{1,3}|^2(t) &= \frac{1}{T_1 - t} \left| \frac{(1 + \delta b)L(t, T_1, T_2)}{b + (1 + \delta b)L(t, T_1, T_2)} \right|^2 \int_t^{T_1} |\gamma_1(s)|^2 ds \\
&= \frac{1}{T_1 - t} \frac{|1 + \delta b|^2 L^2(t, T_1, T_2)}{|2 + \delta b|^2 S^2(t, T_1, T_2)} \int_t^{T_1} |\gamma_1(s)|^2 ds.
\end{aligned}
$$

Exercise 11.2.

(1) We have

$$
P(t, T_1) = P(t, T_2)(1 + \delta L(t, T_1, T_2)), \qquad 0 \le t \le T_1,
$$

hence

$$
\begin{aligned}
dP(t, T_1) &= P(t, T_2)\delta dL(t, T_1, T_2) + (1 + \delta L(t, T_1, T_2))dP(t, T_2) \\
&\quad + \delta dP(t, T_2) \cdot dL(t, T_1, T_2) \\
&= P(t, T_2)\delta\gamma L(t, T_1, T_2)dB_t^{(2)} + (1 + \delta L(t, T_1, T_2))P(t, T_2)(r_t dt + \zeta_2(t)dB_t) \\
&\quad + \delta\gamma L(t, T_1, T_2)P(t, T_2)\zeta_2(t)dt \\
&= P(t, T_2)\delta\gamma L(t, T_1, T_2)dB_t + (1 + \delta L(t, T_1, T_2))P(t, T_2)(r_t dt + \zeta_2(t)dB_t) \\
&= P(t, T_2)(\delta\gamma L(t, T_1, T_2) + \zeta_2(t)(1 + \delta L(t, T_1, T_2)))dB_t \\
&\quad + P(t, T_2)(1 + \delta L(t, T_1, T_2))r_t dt \\
&= \frac{P(t, T_1)}{1 + \delta L(t, T_1, T_2)}(\delta\gamma L(t, T_1, T_2) + \zeta_2(t)(1 + \delta L(t, T_1, T_2)))dB_t \\
&\quad + P(t, T_2)(1 + \delta L(t, T_1, T_2))r_t dt \\
&= P(t, T_1)\left(\frac{\gamma\delta L(t, T_1, T_2)}{1 + \delta L(t, T_1, T_2)} + \zeta_2(t) \right) dB_t + P(t, T_2)(1 + \delta L(t, T_1, T_2))r_t dt \\
&= P(t, T_1)\left(\frac{\gamma\delta L(t, T_1, T_2)}{1 + \delta L(t, T_1, T_2)} + \zeta_2(t) \right) dB_t + r_t P(t, T_1)dt,
\end{aligned}
$$

hence we have

$$
\zeta_1(t) = \frac{\delta\gamma L(t, T_1, T_2)}{1 + \delta L(t, T_1, T_2)} + \zeta_2(t), \qquad 0 \le t \le T_1.
$$

(2) We have

$$
\frac{dL(t, T_1, T_2)}{L(t, T_1, T_2)} = \gamma dB_t^{(2)} = \gamma dB_t - \gamma\zeta_2(t)dt, \qquad 0 \le t \le T_1. \tag{14.8}
$$

(3) Assuming that

$$\frac{dL(s, T_1, T_2)}{L(s, T_1, T_2)} = \gamma dB_s - \gamma \zeta_2(t) ds, \qquad t \le s \le T_1,$$

we get

$$L(s, T_1, T_2) = L(t, T_1, T_2) e^{\gamma(B_s - B_t) - \gamma^2(s-t)/2 - \gamma \zeta_2(t)(s-t)}, \qquad 0 \le t \le s.$$

On the other hand we have $\mathbb{P}_1 = \mathbb{P}$ since $\zeta_1 = 0$, hence $(B_t)_{t \in \mathbb{R}_+}$ is a standard Brownian motion under $\mathbb{P}_1$ and

$$P(t, T_1) \, \mathbb{E}_{\mathbb{P}} \left[(P(T_1, T_2) - K)^+ \big| \mathcal{F}_t \right]$$

$$= P(t, T_1) \, \mathbb{E}_{\mathbb{P}} \left[(P(T_1, T_2) - K)^+ \big| \mathcal{F}_t \right]$$

$$= P(t, T_1)$$

$$\times \mathbb{E}_{\mathbb{P}} \left[((1 + \delta L(t, T_1, T_2) e^{\gamma(B_{T_1} - B_t) - \gamma^2(T_1 - t)/2 - \gamma \zeta_2(t)(T_1 - t)})^{-1} - K)^+ \big| \mathcal{F}_t \right]$$

$$= P(t, T_1) \int_{-\infty}^{\infty} ((1 + \delta L(t, T_1, T_2) e^{\gamma x - \gamma^2(T_1 - t)/2 - \gamma \zeta_2(t)(T_1 - t)})^{-1} - K)^+$$

$$\frac{e^{-x^2/(2(T_1 - t))}}{\sqrt{2(T_1 - t)\pi}} dx.$$

Bibliography

Andersen, L. and Brotherton-Ratcliffe, R. (2005). Extended LIBOR market models with stochastic volatility, *Journal of Computational Finance* **9**, 1.

Bass, L. (October 7, 2007). Brave new world for the equities-shy, Sunday Morning Post, p. 20.

Bielecki, T.-R. and Rutkowski, M. (2002). *Credit risk: modelling, valuation and hedging*, Springer Finance (Springer-Verlag, Berlin), ISBN 3-540-67593-0.

Björk, T. (2004). On the geometry of interest rate models, in *Paris-Princeton Lectures on Mathematical Finance 2003, Lecture Notes in Math.*, Vol. 1847 (Springer, Berlin), pp. 133–215.

Brace, A., Gatarek, D. and Musiela, M. (1997). The market model of interest rate dynamics, *Math. Finance* **7**, 2, pp. 127–155.

Brigo, D. and Mercurio, F. (2006). *Interest rate models—theory and practice*, 2nd edn., Springer Finance (Springer-Verlag, Berlin).

Carmona, R. A. and Tehranchi, M. R. (2006). *Interest rate models: an infinite dimensional stochastic analysis perspective*, Springer Finance (Springer-Verlag, Berlin).

Chen, R.-R., Cheng, X., Fabozzi, F. and Liu, B. (2008). An explicit, multifactor credit default swap pricing model with correlated factors, *Journal of Financial and Quantitative Analysis* **43**, 1, pp. 123–160.

Chen, R.-R. and Huang, J.-Z. (2001). Credit spread bounds and their implications for credit risk modeling, Working paper, Rutgers University and Penn State University.

Chung, K. (2002). *Green, Brown, and probability & Brownian motion on the line* (World Scientific Publishing Co. Inc., River Edge, NJ).

Cox, J. C., Ingersoll, J. E. and Ross, S. A. (1985). A theory of the term structure of interest rates, *Econometrica* **53**, pp. 385–407.

Da Fonseca, J. and Messaoud, M. (2009). LIBOR market in Premia: Bermudan pricer, stochastic volatility and Malliavin calculus, Bankers, Markets & Investors **99**, pp. 44–57.

Da Prato, G. (2004). *Kolmogorov equations for stochastic PDEs*, Advanced Courses in Mathematics. CRM Barcelona (Birkhäuser Verlag, Basel).

De Donno, M. and Pratelli, M. (2005). A theory of stochastic integration for bond markets, *Ann. Appl. Probab.* **15**, 4, pp. 2773–2791.

Dothan, L. (1978). On the term structure of interest rates, *Jour. of Fin. Ec.* **6**, pp. 59–69.

Duffie, D. and Singleton, K. (2003). *Credit risk*, Princeton series in finance (Princeton University Press, Princeton, NJ).

Eberlein, E. and Özkan, F. (2005). The Lévy LIBOR model, *Finance and Stochastics* **9**, pp. 327–348.

Ekeland, I. and Taflin, E. (2005). A theory of bond portfolios, *Ann. Appl. Probab.* **15**, 2, pp. 1260–1305.

Filipović, D. and Teichmann, J. (2004). On the geometry of the term structure of interest rates, *Proc. R. Soc. Lond. Ser. A Math. Phys. Eng. Sci.* **460**, 2041, pp. 129–167.

Fournié, E., Lasry, J., Lebuchoux, J., Lions, P.-L. and Touzi, N. (1999). Applications of Malliavin calculus to Monte Carlo methods in finance, *Finance and Stochastics* **3**, 4, pp. 391–412.

Glasserman, P. and Kou, S. (2003). The term structure of simple forward rates with jump risk, *Math. Finance* **13**, 3, pp. 383–410.

Gourieroux, C. and Sufana, R. (2003). Wishart quadratic term structure models, Working paper.

Guo, X., Jarrow, R. and Menn, C. (2007). A note on Lando's formula and conditional independence, Preprint.

Heath, D., Jarrow, R. and Morton, A. (1992). Bond pricing and the term structure of interest rates: a new methodology, *Econometrica* **60**, pp. 77–105.

Huang, C.-F. (1985). Information structures and viable price systems, *Journal of Mathematical Economics* **14**, pp. 215–240.

Hull, J. and White, A. (1990). Pricing interest rate derivative securities, *The Review of Financial Studies* **3**, pp. 537–592.

Ikeda, N. and Watanabe, S. (1989). *Stochastic Differential Equations and Diffusion Processes* (North-Holland).

Jacod, J. and Protter, P. (2000). *Probability essentials* (Springer-Verlag, Berlin).

James, J. and Webber, N. (2001). *Interest rate modelling, Wileys Series in Financial Engineering*, Vol. XVIII (Cambridge University Press).

Jamshidian, F. (1996). Sorting out swaptions, *Risk* **9**, 3, pp. 59–60.

Kijima, M. (2003). *Stochastic processes with applications to finance* (Chapman & Hall/CRC, Boca Raton, FL).

Lando, D. (1998). On Cox processes and credit risky securities, *Review of Derivative Research* **2**, pp. 99–120.

Malliavin, P. (1997). *Stochastic analysis, Grundlehren der Mathematischen Wissenschaften*, Vol. 313 (Springer-Verlag, Berlin).

Malliavin, P. and Thalmaier, A. (2006). *Stochastic calculus of variations in mathematical finance*, Springer Finance (Springer-Verlag, Berlin).

Mikosch, T. (1998). *Elementary stochastic calculus—with finance in view, Advanced Series on Statistical Science & Applied Probability*, Vol. 6 (World Scientific Publishing Co. Inc., River Edge, NJ).

Milevsky, M. A. and Promislow, S. D. (2001). Mortality derivatives and the option to annuitise, *Insurance Math. Econom.* **29**, 3, pp. 299–318, 4th IME Conference (Barcelona, 2000).

Øksendal, B. (2003). *Stochastic differential equations*, sixth edn., Universitext (Springer-Verlag, Berlin).

Pintoux, C. and Privault, N. (2011). The Dothan pricing model revisited, *Math. Finance* **21**, pp. 355–363.

Piterbarg, V. (2004). A stochastic volatility forward LIBOR model with a term structure of volatility smiles, Preprint.

Pratelli, M. (2008). Generalizations of Merton's mutual fund theorems in infinite-dimensional financial models, in R. Dalang, M. Dozzi and F. Russo (eds.), *Seminar on Stochastic Analysis, Random Fields and Applications (Ascona, 2005), Progress in Probability*, Vol. 59 (Birkhäuser), pp. 511–524.

Privault, N. and Teng, T.-R. (2012). Risk-neutral hedging in bond markets, to appear in Risk and Decision Analysis.

Privault, N. and Wei, X. (2009). Calibration of the LIBOR market model - implementation in PREMIA, Bankers, Markets & Investors **99**, pp. 20–28.

Protter, P. (2001). A partial introduction to financial asset pricing theory, *Stochastic Process. Appl.* **91**, 2, pp. 169–203.

Protter, P. (2005). *Stochastic integration and differential equations, Stochastic Modelling and Applied Probability*, Vol. 21 (Springer-Verlag, Berlin).

Rebonato, R. (1996). *Interest-Rate Option Models* (John Wiley & Sons).

Schoenmakers, J. (2002). Calibration of LIBOR models to caps and swaptions: a way around intrinsic instabilities via parsimonious structures and a collateral market criterion, WIAS Preprint No 740, Berlin.

Schoenmakers, J. (2005). *Robust LIBOR modelling and pricing of derivative products*, Chapman & Hall/CRC Financial Mathematics Series (Chapman & Hall/CRC, Boca Raton, FL).

Vašiček, O. (1977). An equilibrium characterisation of the term structure, *Journal of Financial Economics* **5**, pp. 177–188.

Wu, L. and Zhang, F. (2006). LIBOR market model with stochastic volatility, *Journal of Industrial and Management Optimization* **2**, 2, pp. 199–227.

Yolcu, Y. (2005). One-factor interest rate models: analytic solutions and approximations, Master Thesis, Middle East Technical University.

Index

absence of arbitrage, 23, 40, 73, 101
adapted processes, 3
affine model, 35
annuity numeraire, 121, 129

BGM model, 150
Black-Scholes
  formula, 19, 150
  PDE, 19
bond
  defaultable, 119
  option, 88
  option hedging, 95
  option pricing, 89
  pricing, 39
  pricing PDE, 43, 106, 209
Brownian motion, 1

calibration, 157
call option, 16
cap, 126
cap pricing, 152
caplet, 125
  volatilities, 152
CEV model, 34
CIR model, 33
compounding
  exponential, 56
  linear, 133
conditional expectation, 165
correlation, 164
correlation problem, 101

coupon bond, 120
covariance, 163
credit default, 115
credit default swap, 120
curve fitting, 97

default rate, 117
Delta, 19
deterministic shifts, 100
drifted Brownian motion, 20

exponential distribution, 116
exponential Vasicek model, 35

filtration, 2
forward
  contract, 55, 133
  curves, 68
  measure, 82
  rate, 55
  spot rate, 66, 72
  swap measure, 137
  Vasicek rate, 67

Gaussian
  distribution, 19
  random variable, 164
  vector, 165
Girsanov theorem, 21, 86

hedging strategy, 25

HJM
  condition, 76
  model, 72, 119
Hull-White model, 79, 111

instantaneous forward rate, 58
inverse change of measure, 92
Itô
  formula, 10
  isometry, 4
  process, 11
  stochastic integral, 4

LIBOR
  market, 133
  model, 133
  rates, 134
long rates, 69

Markov
  process, 167
  property, 41, 76, 87
martingale
  continuous time, 166
  discrete time, 166
  method, 23
  property, 40
measurable, 163
Musiela notation, 61, 69

Nelson-Siegel parametrization, 61, 71

payoff function, 15
Poisson process, 117
portfolio
  self-financing, 18, 96

value, 18
predictable
  representation, 24, 26
  simple process, 4
put option, 15

quadratic variation, 8

recovery rate, 119
risk-neutral probability, 23
riskless asset, 17

short rate, 60
stochastic default, 117
stochastic differential equation, 11
stopping time, 117
survival probability, 115
Svensson parametrization, 62, 71
swap, 131
  measure, 137, 146
  rate, 134
swaption, 131, 142, 153
  approximation formula, 156
  hedging, 145
  pricing, 143
  volatilities, 157

tenor structure, 120, 126
two-factor model, 105

Vasicek model, 33

zero-coupon bond, 39

Author Index

Andersen, L. 172

Bass, L. 47
Bielecki, T. ix
Björk, T. ix, 71, 100, 171, 172
Brace, A. 149
Brigo, D. ix, 33, 46, 92, 107
Brotherton-Ratcliffe, R. 172

Carmona, R.A. ix, 33, 171, 172
Chen, R.R. 120
Cheng, X. 120
Chung, K.L. 167
Cox, J.C. 33

Da Fonseca, J. 173
Da Prato, G. 171
De Donno, M. 171
Dothan, L. 34
Duffie, D. 119

Eberlein, E. 172
Ekeland, I. 171

Fabozzi, F. 120
Filipović, D. 171
Fournié, E. 173

Gatarek, D. 149
Glasserman, P. 172
Gourieroux, C. 172
Guo, X. 118, 119

Heath, D. 65, 76
Huang, C.F. 202
Huang, J.Z. 120
Hull, J. 79

Ikeda, N. 4
Ingersoll, J.E. 33

Jacod, J. 163, 165
James, J. ix, 33, 34, 172
Jamshidian, F. 202
Jarrow, R. 65, 76, 118, 119

Kijima, M. ix, 33
Kou, S. 172

Lando, D. 118, 119
Lasry, J. 173
Lebuchoux, J. 173
Lions, P.-L. 173
Liu, B. 120

Malliavin, P. 173
Menn, C. 118, 119
Mercurio, F. ix, 33, 46, 92, 107
Messaoud, M. 173
Mikosch, T. 19
Milevsky, M.A. 173
Morton, A. 65, 76
Musiela, M. 149

Øksendal, B. ix, 19
Özkan, F. 172

Pintoux, C. 34
Piterbarg, V. 172
Pratelli, M. 171
Privault, N. 34, 159, 202, 217
Promislow, S.D. 173
Protter, P. 1, 9, 11, 22, 26, 27, 41,
    43, 87, 163, 165, 202

Rebonato, R. 33, 158
Ross, S.A. 33
Rutkowski, M. ix

Schoenmakers, J. ix, 141, 149, 156,
    157, 159
Singleton, K. 119
Sufana, R. 172

Taflin, E. 171
Tehranchi, M. ix, 33, 171, 172
Teichmann, J. 171
Teng, T.R. 202, 217
Thalmaier, A. 173
Touzi, N. 173

Vašiček, O. 33, 44, 45

Watanabe, S. 4
Webber, N. ix, 33, 34, 172
Wei, X. 159
White, A. 79
Wu, L. 172

Yolcu, Y. 33

Zhang, F. 172